ENDURING
BATTLE

ENDURING
BATTLE
American Soldiers
in Three Wars,
1776–1945

Christopher H. Hamner

UNIVERSITY PRESS OF KANSAS

Published by the University Press of Kansas (Lawrence, Kansas 66045), which was
organized by the Kansas Board of Regents and is operated and funded by Emporia
State University, Fort Hays State University, Kansas State University, Pittsburg State
University, the University of Kansas, and Wichita State University

Library of Congress Cataloging-in-Publication Data

Hamner, Christopher H.
    Enduring battle : American soldiers in three wars, 1776–1945 / Christopher H. Hamner.
    p. cm. — (Modern war studies)
    Includes bibliographical references and index.
    ISBN 978-0-7006-1775-3 (cloth : alk. paper)
    1. Combat—Psychological aspects. 2. Psychology, Military. 3. Soldiers—United
States—Psychology—History. 4. United States—History—Revolution, 1775–1783—
Psychological aspects. 5. United States—History—Civil War, 1861–1865—Psychological
aspects. 6. World War, 1939–1945—Psychological aspects. 7. World War, 1939–1945—
United States. I. Title.
    U22.3.H248 2011
    355.3'30973—dc22

                                                                    2010047205

British Library Cataloguing-in-Publication Data is available.

Printed in the United States of America
10 9 8 7 6 5 4 3 2 1

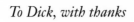

To Dick, with thanks

Contents

Acknowledgments

I am deeply indebted to a number of people who guided me in this effort. William Barney, Alex Roland, Gerhard Weinberg, and the late Don Higginbotham provided generous suggestions on the earliest drafts and enriched the book enormously. Stephen Biddle, Tami Davis Biddle, and Michael Hunt all offered insight and advice along the way, as well as the example of their own stellar scholarship. Peter Kindsvatter and Peter Mansoor graciously read and commented on the penultimate draft of the manuscript, improving it immensely. Michael Briggs and Jennifer Dropkin at the University Press of Kansas exhibited superlative patience while shepherding the project to publication.

Several institutions provided support as well. A fellowship year from Harvard University's John M. Olin Institute for Strategic Studies afforded access to a peerless group of social scientists and contributed immeasurably to the development of these ideas. Support from the U.S. Army's Center of Military History likewise furnished opportunities to engage the evolving arguments with an exacting, accomplished, and supremely knowledgeable group of specialists. I am deeply grateful to both institutions.

George Mason University provided support for the process in a number of ways. I cannot imagine a more welcoming place to undertake a project like this one. Research sabbaticals at two critical junctures helped keep the process on track, and teaching enthusiastic undergraduate and graduate students encouraged me to refine my ideas and approaches along the way. Finally, and most significantly, having genuinely engaged colleagues on the faculty consistently challenged me to expand my approach to the material. Several merit special mention: Michael Bottoms, Meredith Lair, Alison Landsberg, Chuck Lipp, Deborah Kaplan, Matt Karush, and the late Roy Rosenszweig all listened patiently as the ideas here took shape; read and commented on portions of the draft; and offered invaluable advice and cheerleading during revisions.

A prized group of friends deserves thanks for their tireless support. Alex Morss supplied both humor and a twenty-four-hour hotline, generously sharing his vast medical knowledge and explaining the complicated physiology of everything from gunshot trauma to trench foot in simple language. Elizabeth Bridges, Anastasia Christman, Matthew McGill, Edward Schaaf, and Sarah Whelan provided encouragement, much-needed levity, and—most important—friendship and a sense of perspective. Without them, the journey would have been far more difficult and much less enjoyable.

A number of family members deserve special thanks. My aunt, Ann Hamner, watched over me during a decade of research and writing, providing the comfort of hundreds of home-cooked meals, a treasured seat next to her at Tar Heel basketball games, and, most important, wise and honest life advice at every turn. My sister, Eliza Hamner-Koenig, offered not just a sympathetic ear throughout the process but reliable editing insights regarding everything from organization to word choice. It is my good fortune to be related to such a careful and gifted writer, and the finished book is better for her suggestions.

My mother died while I was still a graduate student, before this book was complete. And though I often wonder what a woman who refused to allow her young son to play with toy guns finally made of the fact that I grew up to study the experience of soldiers in combat, there is no question that her influence runs throughout these pages—indeed, through everything I have ever written. Her lifelong dedication to her children's education—beginning with the books she surrounded us with as toddlers—is, I am sure, the single strongest light that led me to pursue a life of letters. Given her love of books, I can only hope that she would be gratified to see that her son wrote one himself and would rightly view the effort as a gesture of profound affection for her.

In the years since my mother's death, my father has done the work of two in providing encouragement. I cannot conceive of a parent offering more unqualified support or exhibiting more patience and understanding toward a son, nor can I find the words to express adequately my love for him and my appreciation for all that he has done. I can only hope that these few lines suggest the magnitude of the debt I owe them both; a full accounting of my gratitude would fill a book of its own.

There is a final debt I owe. Since I first arrived in Chapel Hill, Richard Kohn has set an exceptional standard as a scholar, historian, and mentor. As he has with so many, he helped me become the best historian I could be through gentle encouragement, warm humor, uncompromising standards, and seemingly endless patience. He is quick to downplay the significance of his counsel and example, but his influence is always present in the best parts of my work. My thesis was dedicated to my father, and my dissertation to the memory of my mother. It detracts nothing from their immense contributions to my life that this book is dedicated with gratitude to the person who, more than any other, shaped me as a historian.

Introduction

Writing about his experiences in combat, Civil War veteran David Thompson gave a graphic description of the soldier's essential dilemma under fire. "The truth is," he wrote, "when bullets are whacking against tree trunks and solid shot are cracking skulls like egg shells, the consuming passion in the heart of the average man is to get out of the way." The gruesome violence of the battlefield placed the soldier's natural instincts for self-preservation in direct opposition to the army's equally real insistence that he do his duty. "Between the physical fear of going forward and the moral fear of turning back," Thompson concluded, "there is a predicament of exceptional awkwardness, from which a hidden hole in the ground would be a wonderfully welcome outlet."[1]

That hidden hole, the magical exit from combat, never materialized for most soldiers. Yet despite powerful and often overwhelming reservations, they went forward into the maelstrom of battle. This book is about David Thompson and countless soldiers like him: American infantrymen who fought in the War of Independence, the Civil War, and the Second World War, exposing themselves to enemy fire and performing perilous and sometimes deadly duties amid the terror and trauma of the battlefield. The willingness to face the dangers of battle constitutes one of the most enduring puzzles of human behavior, in part because participation in combat demands behavior that violates so many powerful natural instincts for self-preservation. Army Air Forces psychiatrists Roy Grinker and John Spiegel summarized the problem aptly more than half a century ago in their work on World War II aircrews, *Men under Stress:* "What," they asked, "can possess a rational man to make him act so irrationally?"[2]

Enduring Battle takes this mystery as a starting point. The following chapters search for changes and continuities in the way that soldiers were motivated to face battle from within and without by comparing the experiences of ground soldiers in three different wars.[3] The first

1

chapter explores some of the ways that technology reshaped the experience of infantry combat, beginning with the War of Independence and extending to the Civil War and Second World War, as warfare became progressively more industrialized and technically sophisticated. Subsequent chapters connect those changes to specific ways that soldiers' perceptions of battle and their reactions to it evolved over time.

The focus in this work is mainly on the individual soldier's experience of combat. This approach is something of a departure: to date, the bulk of the discourse on combat motivation has focused on battle primarily as a group phenomenon. The following pages return to a few central questions again and again: What makes soldiers fight? How do they manage to suppress their fears, stifle their instincts for self-preservation, and marshal the will to kill enemy soldiers amid the terror and confusion of the battlefield?

Such questions have bedeviled armies for centuries. Over time, a number of hypotheses have emerged to explain them. Most brutally straightforward is the contention that vast numbers of soldiers fought because they were forced to do so. Military systems furnished direct and uncompromising punishments that persuaded combatants to suppress their fears, at least temporarily, and go forward into enemy fire. Coercion drove soldiers in most eighteenth- and nineteenth-century European armies. Prussian general Frederick the Great believed that an effective infantryman must go into battle "more afraid of his officers than of the dangers to which he is exposed."[4] The power of coercion as a motivator is easy to comprehend: officers and noncommissioned officers replaced the potential danger of enemy bullets with the more direct certainty of friendly ones. (An American Marine in the First World War overheard a terse description of this coercive mechanism: "Don't you turn yellow and try to run," his sergeant shouted at a comrade. "If you do and the Germans don't kill you, I will.")[5] But coercion alone could not always suffice to compel soldiers to fight. In some cases, the practical realities of the battlefield made it difficult to apply punishments effectively and reliably; in other cases, social or cultural expectations prevented armies from motivating soldiers expressly with the lash.[6] When a Northern nurse asked Abraham Lincoln why he did not order more executions in the Union armies given the prevalence of desertion in the ranks, he replied, "Because you cannot order men shot by dozens and twenties. People won't stand it, and they ought not stand it."[7]

Recognition that coercion alone could not furnish a universal explanation for soldiers' willingness to fight gave rise to another hypothesis to explain behaviors observed even in the absence of direct and explicit threats. That explanation posited a class of pressures and inducements that were based not on the promise of physical punishment but instead on the desire for acceptance from some valued social group. Concern for reputation, friendship, mutual interdependence, and trust among members of a unit, according to this argument, could eclipse soldiers' own instincts for survival, if only for a time. Following the Second World War, a trio of influential studies distilled those factors into a single concept: primary group cohesion. Released between 1947 and 1949, S. L. A. Marshall's *Men against Fire*, Edmund Shils and Morris Janowitz's "Cohesion and Disintegration in the Wehrmacht in World War II," and Samuel Stouffer's *The American Soldier* all credited primary group cohesion with the ability to help soldiers endure the physical and mental stress that worked suddenly or over time to break them down and render them incapable of performing their duties.[8] In infantry combat, the primary group usually constituted some small unit whose members knew each other intimately and had regular, face-to-face contact. The combination of affection, obligation, and concern for standing within the group coalesced to create loyalties that overshadowed the individual's concern for his own well-being, at least temporarily. Marshall defined the relationship plainly: "I hold it to be one of the simplest truths of war," he wrote, "that the thing which enables the infantry soldier to keep going . . . is the presence or presumed presence of a comrade."[9]

In the decades following the Second World War, primary group cohesion became the orthodox theory of combat motivation, employed by historians to explain soldiers' often inscrutable behaviors in a variety of eras and conflicts. The appeal of the group cohesion thesis is twofold: it is both consistent with the firsthand observations of countless generations of soldiers and intuitively satisfying. Scholars continue to assign the group cohesion thesis considerable explanatory power. The notion that soldiers fight primarily for one another has appeared repeatedly in the literature on battlefield experience and combat motivation to explain behavior in a wide variety of conflicts, eras, and cultures. Sociologist Nora Kinzer Stewart built her analysis of motivation in the 1982 Falklands Islands War, *Mates and Muchachos*, upon the idea.[10] Civil War historian James McPherson stressed the importance of these bonds in his work on combat motivation among

Union and Confederate soldiers, titling the book *For Cause and Comrades*.[11] In *Why the Vietcong Fought*, political scientist William Darryl Henderson emphasized differences in levels of cohesion to explain the success of the North Vietnamese Army against the larger and better-equipped U.S. Army during the Vietnam conflict. Cohesion and morale, Henderson argued, "were high upon the initial commitment of U.S. units to Vietnam," especially in units that trained together, deployed to Vietnam together, and entered combat together, but they unraveled over time due to U.S. personnel policies; in contrast, the Vietcong were generally able to maintain a high level of cohesion among their fighting units even late in the war.[12] In the introduction to his 1997 book on American GIs in the Second World War, *Citizen Soldiers*, Stephen Ambrose asserted that "unit cohesion, teamwork, the development of a sense of family in the squad and platoon" formed not just the basis of World War II veterans' explanations for their ability to face combat but the basis of his own scholarly work as well: "That is the theme of almost all my writing about the military, from Lewis and Clark to George Armstrong Custer to Eisenhower and D-Day."[13] And in 2003, the Strategic Studies Institute at the U.S. Army's War College released *Why They Fight: Combat Motivation in the Iraq War*, which concluded that "cohesion, or the strong emotional bonds between soldiers, continues to be a critical factor in combat motivation." American soldiers deployed to Iraq, the authors argued, "continue to fight because of the bonds of trust between soldiers."[14] The article provided numerous statements from American soldiers serving there, all interviewed by the authors shortly after their experience in battle, to support that argument. Many stressed the existence of strong emotional bonds among the members of their combat units. "I know that as far as myself," one said, "I take my squad mates' lives more important than my own"; another asserted that the squad was "just like a big family. . . . It is kind of comforting." A third reasoned that these bonds explained his and his comrades' willingness to fight: "It was just looking out for one another. We weren't fighting for anybody else but ourselves." In a statement that could have come directly from Marshall's *Men against Fire* fifty years earlier, the same soldier concluded, "we were just fighting for one another."[15]

The persistence of the group cohesion thesis as the orthodox explanation of combat motivation reflects an assumption that some universal elements of combat (danger, chaos, the risk of death, and the necessity of killing) are so profound that they somehow transcend the influences of technology, tactics, and culture. The belief that combat is, at its core, a universal experience facilitates the belief that there is a single explanation for soldiers' behavior in *all* battle. As historian Richard H. Kohn has observed, "Many scholars have treated battle as a constant—have searched, with little regard for time and place, for the factors that explain why men fight."[16] Among researchers, historians' acceptance of group cohesion as a blanket explanation to account for all motivation in combat is particularly puzzling because it is so fundamentally ahistorical: by definition, a one-size-fits-all explanation ignores the considerable ways that the experience of battle has changed over time. Certainly there has been no shortage of changes in ground combat over the past few centuries: the introduction of firearms, high-powered artillery, automatic weapons, smokeless powder, wireless communications technology, and air power are but a few of the most obvious. Given the easily observed ways in which military technology has changed the battlefield over the last three hundred years, it is surprising that so many scholars expect to find a universal theory to explain the ways that humans have responded to its varied and terrifying rigors. It is clear that battle demanded different things of soldiers in different centuries; scholars should expect to find variations and patterns in the mechanisms that motivated combatants in these different environments.

One other fact is clear: *some* factor or combination of factors has enabled soldiers to overcome their fears long enough to perform their duties in combat. The record of American troops over the past two-and-a-half centuries—from Saratoga to New Orleans, Antietam to the Wilderness, Belleau Wood to Omaha Beach, Chosin Reservoir to Khe Sanh to Fallujah—provides robust evidence that most American soldiers did indeed locate ways to master their anxieties sufficiently to participate in battle. In the main, there have been no mutinies, mass defections, or large-scale surrenders in American armies: the vast majority of troops have stayed and fought when called to.[17] What motivated these soldiers to fight? How and why did their motivations change over time?

Despite its ubiquity, the orthodox explanation cannot explain all of these mysteries. Empirical evidence is often at odds with the hypothesis: in many historical examples, soldiers continued to fight even after battle eroded or destroyed their primary groups. In addition, as experimental psychologists and sociologists have demonstrated, tight primary group cohesion often works against organizational goals, emboldening group members to resist authority imposed from above—precisely the kind of opposition that is devastating in hierarchical organizations like military units. Finally, there is the very strength of the emotional bonds themselves: the group cohesion thesis holds that soldiers are willing to risk their own lives and fight because of the deep bonds of comradeship they feel with their fellow soldiers. But strong emotional ties among soldiers do not necessarily, or even logically, lead to a desire to fight: an infantryman who charges into battle primarily to support a comrade with whom he shares a strong bond of affection may find his attention shifted away from the unit's military objective if, for example, that friend suddenly suffers a combat wound. *Enduring Battle* takes up the mysteries of soldiers' behavior in battle with the expectation that some of their motivations to face enemy fire changed over time, linking the particular requirements of fighting in different eras with factors that helped combatants withstand those specific rigors.

American infantry armies provide an ideal arena for this study. The infantry does most of the fighting, if not most of the killing, and as a result takes most of the casualties in battle (historian Russell Weigley has estimated that the average World War II U.S. infantry regiment lost a hundred percent of its personnel to battlefield casualties in just three months of fighting).[18] The demands placed upon infantry troops in combat are as high as those placed upon any soldiers: historically, infantry armies have faced the danger of battle at both the sharpest and most vulnerable point. Infantry combat in each of these three wars involved the kind of individual and small-group dynamics found in few other types of combat (members of a tank or bomber crew, for example, are bound together in concrete ways and lack the flexibility of decision making that infantrymen usually possess). For a variety of reasons, American forces have used physical coercion to a much lesser extent than most other armies.[19] And much of the best and most expansive work on combat motivation in history and social science focuses upon American armies: the United States has been unusually willing to examine its soldiers' behaviors openly and critically.

Enduring Battle explores both the changes and continuities in infantrymen's combat experience over time. Important commonalities connected the experiences of soldiers on the battlefields of these wars: the smothering presence of danger, the potency of fear in battle, and the constant proximity of death represented the most profound and nearly unchanging elements of the soldier's lot in battle. Scholars who stress the universal nature of combat point to precisely these factors to substantiate their case. But these continuities, though intense, represent only a part of the story: the particulars of infantry combat changed over time, often dramatically. A reader who skims accounts of infantry combat from different centuries is struck quickly by the differences in the details of the descriptions. Veteran infantry soldiers recall some elements of the experience that are similar regardless of time and place (the fear of death and injury, and the overpowering chaos and confusion of the battlefield, surface in nearly every account of infantry combat, no matter its origins), but even a casual overview of accounts from different periods reveals important contrasts as well. Some of those differences are in the trappings of battle, but others are more substantive. A reader confronted with a young GI's account of being strafed by German aircraft during the 1942 North African campaign would not easily mistake that description for a Confederate rifleman's portrayal of charging the enemy line at the 1862 battle of Antietam, even though both narratives revolve around the presence of danger and death. What those soldiers saw and did, and the way they registered their experiences, varied in important ways.

Many of those differences stem from the immense technological changes in the hardware of war that occurred in the intervening decades. Together, those innovations wrought a staggering number of changes in warfare. But from the point of view of the individual soldier caught in the maelstrom of battle, the broad range of evolutions that transformed infantry combat from the dawn of the gunpowder age to the twentieth century coalesced into a few obvious tactical differences. In the seventeenth, eighteenth, and nineteenth centuries, infantry troops marched into combat in dressed lines: individual soldiers deployed shoulder-to-shoulder alongside their fellows in configurations descended directly from the close-order formations of the Greek phalanx. Those linear formations helped mass the fire of the muskets troops carried, compensated in part for the weapons' limited accuracy, and simplified officers' command over their troops. They also maximized officers' ability to communicate with their soldiers and facilitated the use of banners, bugles, brightly colored uniforms, and shouted commands to reduce the inherent confusion of the battlefield. Perhaps most importantly, the close-ordered ranks of the linear system made it easier for officers and noncommissioned officers to police their troops' behavior and to punish those whose nerve appeared to waver. Eighteenth- and nineteenth-century armies arrayed their troops in lines because few viable alternatives existed.

By the middle of the twentieth century, the way that infantry armies deployed in battle had changed dramatically. The introduction in the latter decades of the nineteenth century of vastly more accurate rifled muskets, quicker-firing magazine-fed automatic weapons, and high-powered, long-range artillery forced a profound change in the way armies arrayed their infantry soldiers in combat. Together, the innovations in arms technologies created an order-of-magnitude increase in the number of projectiles in the air, a development that one German soldier of the First World War referred to vividly as the "storm of steel."[20] The massive increase in the amount (not to mention the accuracy) of metal flung at infantry soldiers rendered the massed, shoulder-to-shoulder troop dispositions of the linear battlefield suicidally obsolete. Slow-moving, exposed formations could be raked by automatic fire and targeted by long-range artillery bursts with ruinous effect, as the first British waves at the Somme in 1916 illustrated so tragically. So profound were the technical improvements in warfare made possible by industrialization that they led Polish political theorist Jan Bloch to conclude at the close of the nineteenth century that warfare had become too terrible to wage: science and industry had finally created a battlefield that citizen-soldiers (who would necessarily constitute the armies of future European wars) would simply be unable to bear. The supreme violence of industrialized combat, Bloch argued in one of his essays, eventually would render warfare all but impossible by turning the battlefield into a place that no reasonable person could face.[21]

As the conflicts of the twentieth century demonstrated so clearly, the new weapons did not put an end to war. After some immensely costly experiments during the First World War, planners ultimately discovered ways to mitigate the effects of the storm of steel. The solution to the problem was one first developed during the Boer War of the late nineteenth century: dispersing infantry soldiers from their lines. If automatic weapons and increasingly devastating artillery blasts made tightly packed dressed ranks too vulnerable to operate effectively, then infantry would simply no longer fight from those traditional formations. Dispersed tactics (sometimes referred to as the "modern system of tactics," since they represent virtually the only way for an army to operate against the staggering destructive power of modern weaponry) scattered soldiers so that no single machine gun or artillery blast could annihilate large numbers. Soldiers in the dispersed system employed cover (interposing an obstacle like a rock, tree, or dip in the terrain to deny the defender a clear line of sight) and concealment (using camouflage to deny the defender an obvious target) to provide adequate if temporary refuge from the dangers of the storm of enemy projectiles, and they coordinated their movements with their comrades (one element scrambling for cover while another provided suppressive fire that forced opposing soldiers to keep their heads down, frustrating the defenders' ability to return fire). As the successful German offensives at the end of World War I demonstrated, dispersed tactics and the use of cover and concealment allowed infantry armies to once again take the initiative even in the face of what had once seemed overwhelming defensive firepower.[22] The advent of the "empty battlefield" of the Second World War (without the large formations, brightly colored uniforms, conspicuous banners, flags, music, and officers that so distinctly characterized the close-order linear battlefield) was a direct result of these tactical evolutions.

While dispersion provided infantry soldiers adequate opportunity to operate effectively without being obliterated immediately by enemy fire, the nature of the tactics themselves created new problems. Soldiers scattered across the battlefield could no longer be monitored directly, and officers and noncommissioned officers (NCOs) could not make credible coercive threats to soldiers they could not keep watch over constantly. Open-order tactics also broke down individuals' ability to police one another; the French soldier and military theorist Charles Ardant du Picq captured the nub of this part of the problem late in the nineteenth century when he observed that on the dispersed battlefield, "unity is no longer assured by mutual surveillance."[23] And the necessity of dispersion, and the use of cover and concealment, created new opportunities for soldiers looking to dodge the danger of battle. As William Slim, commander of the British troops during the Second World War's Burmese campaign, asserted, "Nothing is easier in jungle or dispersed fighting than for a man to shirk." If the soldier has "no stomach for advancing," Slim held, "all he has to do is flop into the undergrowth; in retreat, he can slink out of the rear guard, join up later, and swear he was last to leave."[24] Mechanisms motivating soldiers in combat that had proved relatively clear on the close-order linear battlefield, where every move was observed by comrades, could not function with the same effect on the empty battlefield of the mid-twentieth century. Military planners confronting the technical necessity of abandoning linear tactics worried, not unreasonably, that they would simultaneously forfeit the critical tools that historically had enabled armies to keep soldiers fighting: threats of physical coercion, close monitoring, policing by comrades, and direct control.

As the record of infantry combat from the mid-twentieth century on indicates, military systems ultimately overcame those challenges to motivating soldiers, just as they had overcome the tactical challenges presented by the storm of steel. The comparative examination of the experience of battle provides a way to examine exactly how combat and combat motivations changed over time. It is almost impossible to see many of the most interesting changes in infantry combat in isolation. Viewing one battle or one war emphasizes the features that helped motivate a particular set of soldiers, but offers few insights into the nature of the changes over time or the reasons for those changes.

Juxtaposing individual experiences of infantry combat from three such distinctive conflicts throws the differences into sharp relief. Four in particular stand out. First, over time, battle for the individual soldier became increasingly depersonalized, as the increasing range and power of weapons removed first the soldier's enemies and ultimately most of his comrades from view. Depersonalization increased as the distance between opponents grew. Combat that had once been a hand-to-hand melee with a visible and concrete enemy gradually gave way to lines firing at 30 yards, then 100 yards, then a quarter-mile, until by the mid-twentieth century soldiers found themselves targeted by artillery miles away and from aircraft hundreds or thousands of feet overhead. The growing depersonalization of combat bred acute and sometimes debilitating frustration among those exposed to it, as individual soldiers felt more and more powerless to respond to a distant enemy they could not see.

In addition, battle came to feel more isolated, as soldiers found themselves more and more alone, without the reassuring presence of comrades. Dispersed tactics heightened feelings of isolation: as formations spread out, individuals lost the physical comfort that came from the tangible presence of fellows in the line. That "touch of the elbow," so critical to steeling the wills of Civil War soldiers to fight, dissolved on the battlefields of the mid-twentieth century. Nothing but a stark sense of isolation filled the void. Where it had once been nearly impossible for a soldier to feel isolated in combat—battles of the eighteenth and nineteenth centuries resembled two armed mobs crashing into one another—it was almost equally impossible for a soldier to avoid a frightening isolation on the dispersed battlefield at times. Desperation was a common byproduct of that isolation: as S. L. A. Marshall himself wrote of the sensations of being shelled by mortars while separated from his unit, "The terror I knew was almost overwhelming." In Marshall's estimation, that feeling was not an aberration but the rule: "It always happens that way. Be a man ever so accustomed to fire, experiencing it when he is alone and unobserved produces shock that is indescribable."[25]

As technological developments made weapons more powerful and more accurate, combat came to feel more specific to the individual soldier caught in its grasp. Standing within the orderly lines of a Continental infantry regiment weathering an enemy volley during the War of Independence, few soldiers registered the sensation of being targeted specifically. Among close-ordered ranks exposed to unaimed volley fire, who was hit and who was not appeared to be mainly a matter of chance. The far more accurate weapons of the twentieth century, and the corresponding rise of aimed fire, changed the way that battle felt to the targets of enemy bullets. The object of that aimed fire frequently felt that he was being targeted, personally and specifically, by the projectiles. For the World War II soldier cowering in his foxhole and bracketed by opposing artillery (the first shell falling ahead of the position, a second behind, and a third directly on the target as the opposing crew found the range), the experience created the terrifying sensation of an enemy groping for him specifically, rather than simply being the unlucky recipient of shells lobbed indiscriminately into a massed formation.

A final change occurred in degree rather than in kind. As warfare became more technologically sophisticated and increasingly industrialized, the length of battles, and the amount of time individuals were exposed to enemy fire, increased steadily. Battle in the eighteenth century might last less than an hour; by the time of the Civil War, multiday battles were common, though darkness and poor weather still forced a lull, and combat was more sporadic, with long periods in camp or on the march separating the terrifying episodes. Night, poor weather, and bivouac offered soldiers of the Second World War no such respite: battle continued around the clock, so that an individual soldier might be under fire (or at least under the threat of fire) for days or weeks at a time.

Together, those factors changed the experience of fighting for the individual soldier in some profound ways. As infantry combat moved from close-order linear tactics to open-order dispersed tactics, the nature of the individual soldier's role in battle had to evolve accordingly. Where the effective soldier on the linear battlefield had to be an automaton, the effective soldier on the dispersed battlefield had to be autonomous. On the linear battlefield, the infantry soldier's challenge was to stanch visceral, instinctive fears and to weather the weight of enemy fire while executing the straightforward tasks (loading and firing his weapon, for example, or moving with the company from marching column to firing line) that knit the unit together and enabled it to answer with its own volleys. Individual initiative had little value on the battlefields of the War of Independence and the Civil War; indeed, soldiers exercising autonomy (deciding independently whether to seek cover or charge an enemy position) threatened disaster, risking the integrity of the entire formation and eroding its ability to maneuver and return fire effectively. On the linear battlefield, a unit's success depended in large part upon the stoicism of the individual soldiers who constituted it. The critical importance of weathering enemy barrages without succumbing to the overwhelming urge to take cover was apparent even to the lowliest soldiers functioning within the linear system: Private David Perry, an American colonist who fought in the ranks shoulder-to-shoulder alongside British regulars at Ticonderoga in 1759, remembered that this insight into the nature of close-order combat "greatly surprised me, to think that I must stand still to be shot at."[26]

New tactics placed new demands upon combatants. The dispersed battlefield and the storm of steel rendered useless soldiers acting as automatons, attempting to brave enemy fire in the open. Battle had become far too complex, and technology far too destructive, for naked stoicism to be useful on its own. Rather than simply mustering the will to stand still while being shot at, the effective soldier on the dispersed battlefield had to exercise an enormous amount of individual initiative (where to seek cover, when to sprint for the next refuge, where to aim suppressing fire) in order to achieve a military objective. Training in the modern system of tactics could teach soldiers the tasks they ought to perform in a given situation. But the very nature of the dispersed battlefield eliminated the possibility of constant surveillance by officers and NCOs, and removed the traditional guarantee that soldiers would expose themselves to enemy fire and put their training into action. For the military system, the challenge was to imbue soldiers with the kind of autonomy necessary to function on the dispersed battlefield while simultaneously ensuring that those soldiers did not exercise that autonomy to flop into a shellhole, feign a sprained ankle, or defect from combat in any of the other scores of ways available to a soldier on the dispersed battlefield.

This is hardly the first work to recognize the changing reality of combat for the individual soldier. John Keegan's *The Face of Battle* juxtaposed three vastly different battles (the 1415 battle of Agincourt, the 1815 battle of Waterloo, and the 1916 battle on the Somme) to reconstruct the British soldier's experience of fighting in different eras from the bottom up. Examining the reasons those men went forward, Keegan concluded that they fought in large part out of the necessity of survival and self-interest. Once the military system delivered troops to the battlefield, the individual's safest course of action was often committed participation in the fighting. Acting otherwise risked forfeiting the safety of the group and hazarding the danger of isolation, which often exposed the soldier to greater danger than the firing line itself. The power and reality of combat's "unavoidable circumstance," as Keegan termed it, channeled many soldiers' efforts into the fight. Going forward offered at least the possibility of neutralizing the enemy and thus ending, or at least blunting, the threats directed against the soldier.[27] Keegan's findings echoed in a more comprehensive work on battlefield behavior, Anthony Kellett's synthetic monograph *Combat Motivation*. Like Keegan, Kellett concluded that most soldiers fought when required to do so: "usually without notable enthusiasm, but equally without widespread or persistent defection." In Kellett's estimation, the willingness of the majority to fight stemmed in large part from pragmatic factors. The military system successfully maneuvered soldiers to the firing line, where few alternatives to fighting existed and where, in Kellett's words, "the penalties of not fighting (personal and social as well as disciplinary) outweighed the uncertain risks of fighting."[28]

The close examination of soldiers' experiences and reflections from these three conflicts argues that armies in both the linear and dispersed systems employed many common techniques, as Kellett suggested, to get their soldiers to fight. But the relative emphasis of those techniques shifted in response to technological and tactical evolutions on the battlefield. Within the linear system, the emphasis fell on strengthening the *known* penalties of *not fighting*, particularly by enforcing direct physical and social punishments for soldiers reluctant to go forward. Within the dispersed system, relatively more weight fell on shading the soldier's interpretation of the *uncertain* risks of *fighting* itself: increasingly, soldiers acquired reasons to believe that active participation in combat (within the military's carefully constructed framework) would maximize their chances for survival and diminish the chances of being killed or wounded. The story of how those transitions occurred is woven throughout the chapters of this book.

As becomes clear in the following pages, the deceptively simple question "What makes soldiers fight?" has no equally straightforward answer. A variety of factors, from training to leadership to the soldier's relationship with his arms and with his comrades, all affected the ways that infantrymen prepared themselves for battle and were motivated from without and from within to endure its rigors. The most importance influence of technology over time was to alter the relative importance of those different factors and to transform the ways they were implemented and applied. Within the linear system, individual soldiers depended on a host of motivators to help them endure the horror of standing exposed and absorbing the weight of enemy volleys: direct monitoring and coercive threats, the herd instinct and the tangible bodily reassurance of others, deeply ingrained physical manipulations, and the near presence of leaders and fellow soldiers whose behaviors could be imitated. The combat record of the Continental Army and of both Civil War armies suggest that this collection of motivators was extremely effective in getting soldiers to fight. But the plain realities of the dispersed system rendered the power of many of those motivators diffuse. Spread-out soldiers could no longer feel the tangible presence of their comrades, and coercive threats were more difficult to make and enforce when soldiers were not under mutual surveillance. Too often leaders and comrades were out of visible contact, depriving soldiers of conspicuous neighbors whose actions could be copied. The trauma of the battlefield was equally acute, though the challenge was no longer simply to stand, exposed and stoic, to withering fire. How did those motivators evolve to compensate for the new realities of battle?

One thing that did not change over time was the individual's overriding desire to stay alive. Those sentiments are easy to locate in soldiers' accounts of battle in every era. One World War II GI wrote that the physical and emotional stress of battle put troops under tremendous pressure: exposed to the trauma of combat, he argued, every soldier "regresses to the primal instinct of survival at all cost."[29] Another acknowledged that the individual's overpowering instinct was for self-preservation: "I realized that our bunch of GIs was not fighting for mother, country, and apple pie. Bullshit. We wanted to live."[30] The nearly overwhelming power of the survival instinct posed the largest obstacle to motivating soldiers in every era, and proved especially difficult on the dispersed battlefield, where opportunities to shirk duty were far more available. Of course, to most soldiers, the individual goal of remaining alive appeared directly at odds with the goals of their military system. Indeed, for many, it was precisely the pursuit of their own army's objectives that constituted the most immediate obstacle to survival. (One infantry scout serving in Salerno during the Second World War's Italian campaign put that fatalism into words, reporting that most of the men in his division had given up all hope of being relieved under the assumption that "the Army intended to keep them in action until everybody was killed.")[31] Soldiers, in the main, wanted to stay alive; the army forced them to stand in harm's way.

One of the central arguments in this book deals with the way that military systems encouraged soldiers to fight on the empty battlefield despite those profound reservations. Closing the apparent gap between going forward and the individual's survival instinct helped make active participation in combat seem less like a death sentence: in effect, successful military systems attempted to transform committed fighting into a rational act. As the following pages suggest, that gap was narrowed by hundreds of subtle lessons and suggestions that attempted to align the individual soldier's desire to stay alive with the military's own inflexible demands. One World War II poster depicted a careless GI picking up a German helmet as a souvenir and enveloped in an explosion beneath the legend "Basic Training Counts!" The message below clarified the soldier's fate: "HE missed his basic in BOOBY TRAPS." The explicit message of the poster was as simple as one emblazoned on a flier reminding soldiers of the life-and-death necessity of digging foxholes as a matter of instinct: "DIG! . . . OR DIE!" with a similar appeal to new soldiers' fears and desire to remain alive. But both messages contained a subtler but equally powerful suggestion about the nature of military training itself: *there are things you can learn to do that will save your life.* That secondary message encouraged viewers to think about combat as an environment in which logical cause-and-effect still held, in which individual decisions and actions had consequences: doing *this* causes *that* to happen. It was a sentiment most recruits already wanted to believe: the poster's doomed soldier met his untimely death not because he was unlucky, or because the odds of survival in the meat grinder of twentieth-century industrialized combat were ridiculously slim, but because he was careless and had not paid enough attention to the lessons of his training. Like World War II soldier Geddes Mumford, who reassured his parents from training camp that "most men get killed in battle because they forget to take cover or make some tactical mistake," many soldiers assigned this belief power precisely because it implied a high degree of control over their outcomes in combat. Firm in that logic, they could remain assured, as Mumford did, that survival was likely because "I'll make no mistake like that."[32] The comforting corollary such messages offered new recruits: *you* can learn to avoid booby traps and dig foxholes—*you* won't be killed. Of course, the emphasis on what the soldier could control, and what the individual could do to maximize the chances for survival ("the uncertain penalties of fighting itself," in Kellett's words), could not hold indefinitely. Given enough exposure to the unalterably lethal, pernicious, and random nature of the battlefield, nearly every soldier ultimately arrived at the same conclusion. Despite everything that training, officers, and scores of pamphlets held, battle was at its core horribly and unchangeably arbitrary. Fatalism and breakdown set in as soldiers lost the sense that they could exert some control, however small, over their fates.

That observation—that a perceived alignment of the soldier's natural instinct for self-preservation and the military's goals took on more and more importance as technology forced soldiers to disperse—forms one of the central arguments of this book. That interpretation unfolds in six chapters. The first, "The Evolving Character of Infantry Combat," begins describing the nature of ground combat in different technological eras by outlining some salient features of the infantry soldier's experience of battle in the War of Independence, the Civil War, and the Second World War. The distinctions between the particular dangers experienced by eighteenth-, nineteenth-, and twentieth-century soldiers (the nature of those threats, how they announced themselves, their detectable effects, and soldiers' understanding of them) highlight the ways that the experience of battle both changed and remained the same with the introduction of more and more advanced technologies. Those changes and continuities provide a foundation for the second chapter's examination of the chief obstacle to performing in battle, fear, and its analysis of the ways that troops' experience of that emotion and the ways that the techniques for managing fear evolved over time.

The following four chapters examine some of the ways that those changes played out in four specific areas: in training, in leadership, in the soldier's relationship with his weapons, and in the soldier's relationship with comrades in the ranks. The third and fourth chapters work in tandem, exploring the ways in which military systems have, over time, attempted to deal with the problem of soldiers' fear in battle. The discussion in chapter 3, "Training," focuses on the effects of infantry training as an inoculation against the effects of fear in battle, examining how soldiers absorbed different lessons from training (in obvious and in more subtle ways) that helped influence their behavior in combat before they arrived on the battlefield. Chapter 4, "Leadership," examines the changing patterns of battlefield leadership in the three conflicts. Central to that discussion is the contrast between leadership in the linear and dispersed tactical systems: the behaviors promoted, the values encouraged, and, most importantly, soldiers' changing ideas about the nature of leadership itself.

Two subsequent chapters discuss soldiers' experiences in battle and the relationships that informed and shaped those experiences. Chapter 5, "Weaponry," examines soldiers' changing relationship with their weapons, as the tools of war evolved from the clumsy, slow-firing shoulder arms of the eighteenth century to the automatic weapons of the twentieth, and how soldiers' understanding of their weapons affected their willingness to fight. The final chapter, "Comradeship," examines the ways that soldiers' interactions with friendly soldiers evolved over time, comparing the mechanics of these interactions in both the linear and dispersed systems and highlighting the ways in which the "band of brothers" phenomenon functioned differently in each—an analysis that suggests a more pragmatic lens through which to understand the evolving power of those bonds to motivate soldiers on the vastly different battlefields of the War of Independence, the Civil War, and the Second World War.

The focus on perception and motivation in battle leads the analysis in these pages away from one significant debate surrounding soldiers' behavior: how important a role does ideology play in motivating soldiers to fight? That question animated some of the most important scholarship on combat motivation in the last half of the twentieth century. Mid-twentieth-century work on World War II soldiers, for example, suggested that those American troops were more cynical, and therefore less likely to fight out of dedication to a patriotic cause, than their predecessors in the American armies of the eighteenth and nineteenth centuries. (When quizzed about what they were fighting for, a majority of World War II GIs reported that they were fighting to get the job done or to get home, rather than to defend the nation or to save democracy, a reality typified by the popular sentiment that the soldiers were fighting for "blueberry pie.")[33] Subsequent scholars of combat motivation employed findings about the motivations of twentieth-century soldiers to cast a retroactively skeptical eye on the sometimes flowery patriotic language of Civil War soldiers: McPherson's book on the motivations of Civil War soldiers, *For Cause and Comrades*, was in part a rebuttal to the argument that the patriotic language of those nineteenth-century soldiers was employed insincerely. Rather than privilege or discount the role of ideology in driving soldiers to war, the research in *Enduring Battle* suggests that behavior in battle is circumscribed by the urgent necessity of combat. The pressures and confusion of ground combat are so intense that ideology is simply disconnected from behavior when the bullets and shrapnel are flying. Lewis Hosea, a Union officer who fought at Shiloh, described this rush as the "frenzy of battle" and wrote that it "seizes upon every other faculty, physical and mental, and makes one oblivious to all other surroundings."[34] Ideology can be an extremely potent force leading soldiers to join the army (it was certainly present among many soldiers in the three conflicts under discussion here, perhaps the quintessential ideological wars of American history), and it can be an equally powerful reason for soldiers to stay with the army so that military institutions can deliver them to the battlefield. Once in battle, however, soldiers found their options dramatically circumscribed by Keegan's "unavoidable circumstance" of combat. Training, conditioning, leadership, and the other factors discussed in the following pages exerted far more pronounced effects on individual soldiers' behavior and life-and-death decisions under fire than could the comparatively abstract and detached influence of political ideology.

A final note on writing about combat: it is a challenge for a scholar who has not served in the military, much less been in harm's way, to write with authority about the soldier's experience of battle. Combat, according to many veterans, is fundamentally ineffable: the sights, sounds, and sensations are not easily captured in words, and the emotions and reactions that battle produces do not easily fit into language. Indeed, the basic act of writing about battle, flattening the violent, confusing, incomprehensible experience to the linear narrative of the printed page, seems paradoxical. As one soldier of the Civil War wrote, "No man can give any idea of a battle by description nor by painting."[35] A World War II GI frustrated by his inability to capture his battle experiences in letters home likened the "stale flatness" of those written missives to a "snapshot taken of a skidding, roaring car passing on a track at high speed. The print, returned a week later, conveys no snarl, no roar, no smell of exhaust, oil, and rubber, no sense of blinding speed. When you display the blurred picture, you can only mumble to the unimpressed viewer: 'Hey, you had to be there!'"[36] Similarly, the World War II veteran and literary scholar Samuel Hynes has suggested that readings soldiers' accounts, whether of Waterloo, the Argonne, or Hué, forces us to "see with estranged eyes." The battlefield, he notes, is utterly unlike anything in the civilian world; it is "not a place we could travel to."[37]

The challenge of reducing battle's confusion and terror to the stark black and white of words on a page, however, applies to all those who attempt to write about battle, regardless of personal experience. More daunting to the scholar untested by combat is the widespread sentiment among veterans that those who have never experienced battle cannot truly understand it. After years and years of reading soldiers' letters, diaries, and memoirs—many unnervingly intimate glances into the minds of human beings attempting to make sense of some of the most traumatic moments of their lives—I have no doubt that this notion is in many ways true. Those spared from battle cannot ever hope to understand the experience and its effects in their terrifying fullness. In a related way, no single soldier can understand *every* experience of battle. The GIs of the Second World War may be able to approach more closely the experiences of their predecessors in the Continental Army of the War of Independence by dint of their common exposure to danger and their shared sacrifices, but even those twentieth-century veterans are distanced in important ways from the experience of combat among the eighteenth century's dressed ranks and musket volleys. We are all separated to some degree from the experiences described in the following pages. Most of us, mercifully, have been spared all of them; a few have experienced some of them. Nevertheless, despite this necessary separation, it is my earnest hope here that the diligent scholar can offer some worthwhile insights into the nature of the experience and the ways it has changed over time.

1

The Evolving Character
of Infantry Combat

Roscoe Blunt went to war in 1944. Inducted into the U.S. Army mid-way through World War II, Blunt spent months training stateside before journeying with the American 84th Infantry Division through France and Holland and then into Germany along with the Allied units that advanced in the push toward Berlin. In 1945 he reached the German village of Lindern, still strewn with wreckage from the ferocious battle recently fought there. Though surrounded by the evidence of some of the most advanced military technologies of the mid-twentieth century (numerous unmanned German V-1 "buzz bombs" had recently passed overhead during the battle, and Blunt himself was strafed by a low-flying Me-109 fighter aircraft as he reconnoitered), he found that the scene of the wrecked battlefield reminded him of nothing so much as Mathew Brady's famous nineteenth-century images of rebel and Union bodies scattered across the fields of the American Civil War. He drew a universal lesson from the strange familiarity: "I suppose any combat veteran would say that one war is the same as any other, that only the uniforms change."[1]

Blunt's observation, that for the individual soldier it was the trappings of battle rather than its substance that changed, was hardly unique. Generations of soldiers gone to war often have remarked on the seemingly universal dimensions of the infantry soldier's experience. In the minds of countless veterans, some essential part of their exposure to the lethal dangers of battle connected them to warriors gone before. One World War II paratrooper, ruminating on the likelihood of his death in coming battle, found a degree of solace in the link. "As things now looked, I would probably cash in my chips in battle and dissolve into soil and air in some place no one in the states had ever heard of. But what the hell!" he concluded. "Men had been dying like that" since Homer's time.[2]

20

The undeniably life altering effects of facing enemy bullets in combat have struck many infantry veterans as the core of their experience in battle. Soldiers commonly reflected on the transformative power of those experiences, linking themselves to troops long gone: soldiers who fought in different wars, against different enemies, with different weapons, in the service of different causes. In the twentieth century, the American military system sometimes encouraged soldiers to make those connections, as when it sent a group of World War II soldiers on a tour of the Civil War's Bull Run battlefields as part of their training; the soldiers studied terrain problems and listened to a lecture on the battles delivered by an expert furnished by the army.[3] Another World War II GI recalled a new captain's address to the company in which the officer outlined his expectation that the men would "serve as well as our fathers and grandfathers did."[4] Such deliberate connections carried reassuring overtones; offering those links to soldiers fearful of the unknown encouraged the conclusion that they, too, would be able to function in combat, just as other frightened young men in earlier American armies had faced both their own fears and enemy bullets. Equally often, soldiers made the connections between their experiences and those of earlier generations of soldiers independently. One World War II GI, whose experience in the mechanized twentieth-century U.S. Army involved considerably more marching than he had anticipated ("twenty to thirty long, exhausting, dreary miles per day"), reached into the past for a fitting comparison: "Stonewall Jackson's boys in the Shenandoah Valley in Virginia had nothing on us."[5] Civil War veteran Oliver Wendell Holmes, Jr., first contemplated war while studying a picture of an old soldier from the War of Independence ("a white-haired man with his flint-lock slung across his back") and, failing to take into account the passage of years between the veteran's service and the time the image was made, imagined that wars were fought by old men.[6] Anthony Swofford's Marine Recon unit spent their time before deploying in the 1991 Gulf War watching Vietnam films like *Apocalypse Now* to prepare themselves for the unknowns of battle.[7] New generations of soldiers naturally looked back to their predecessors for reassurance and inspiration. Those old soldiers provided encouragement to nervous warriors and a way to measure their own experiences.

While the sentiment behind Roscoe Blunt's observation that "one war is the same as any other" for the individual foot soldier is hardly uncommon among veterans, its accuracy is more difficult to assess. Many scholars have detected an essential similarity among the experiences of those who have faced the trauma of combat. In *The Face of Battle*, historian John Keegan provided a veritable catalog of those commonalities in his examination of warfare from the foot soldier's perspective at the battles of Agincourt, Waterloo, and the Somme. "What battles have in common is human," Keegan wrote. The study of battle is at bottom a study of the "behavior of men struggling to reconcile their instinct for self-preservation, their sense of honor and the achievement of some aim over which other men are ready to kill them." Battle, in Keegan's view, was "always therefore a study of fear and usually of courage; always of leadership, usually of obedience; always of compulsion, sometimes of insubordination; always of anxiety, sometimes of elation or catharsis; always of uncertainty and doubt, misinformation and misapprehension, usually of faith and sometimes of vision; always of violence, sometimes also of cruelty, self-sacrifice, compassion."[8] Nineteenth-century French military theorist Charles Ardant du Picq located the common thread in warfare as the human element that ran through it. Though war "is subject to many modifications by industrial and scientific progress," he wrote, "one thing does not change, the heart of man."[9] Half a century later, a 1941 U.S. Army Field Manual echoed that observation, arguing that "man is the fundamental instrument in war; other instruments change but he remains relatively constant."[10]

Those common areas in the human experience of battle are, without question, very real. Yet there also exist profound variations in the battle experiences of infantry soldiers across time. The nature of infantry combat was very different for the Continental soldier at Saratoga, the American regular at Chapultepec, the rebel infantryman at Chancellorsville, the GI at the Battle of the Bulge, and the Marine at the Chosin Reservoir. What they saw, heard, and felt, and what the situation required them to do and to endure, were similar in the broadest sense but very different in the particulars. And those particulars matter: rather than simply reenacting the same experience over and over again in different uniforms, soldiers of different eras reacted to an enormous range of dangers, cues, and imperatives. Their perceptions and responses reveal a great deal about the evolving nature of soldiering in different eras, particularly in the way individuals mustered the ability to stay and fight even as powerful instincts and rational fears urged them to flee.

Both the changes and continuities are useful in understanding what soldiers underwent in battle. Indeed, many of the most obvious features of these soldiers' experiences appear among the things that did not change. Keegan's catalog of the commonalities of battle is no exaggeration; the experiences of soldiers in combat contain a number of striking features that persist independent of the specific time and place in which a given battle occurred. Even a cursory overview of soldiers' combat experiences in different wars reveals a set of common themes, including the conspicuous danger of the battlefield, the omnipresent nature of death and destruction, the palpable sensations of fear, and the unavoidable necessity of staying and fighting. All these themes emerge from soldiers' recollections of infantry combat, whether the scenes occurred in 1776, 1812, 1863, 1944, or 2003.

Because of their acutely graphic nature, many of the commonalities are especially easy to spot. The overwhelming presence of danger, in particular, fairly leaps from veterans' recollections, no matter when or where they took place. Samuel Webb, for instance, was a twenty-two-year-old lieutenant in the summer of 1775 when the British attacked colonial fortifications on Bunker Hill and Breed's Hill outside Boston. Two days after the battle, Lieutenant Webb described the first sensations of combat in a letter to his brother: "Cannon & Musket Balls," he wrote, "were flying about our Ears like hail & a hotter fire you can have no idea of." The weight of the British artillery bombardment left a distinct impression on Lieutenant Webb, who noted that enemy shot "flew as thick as Haile Stones" as he approached the battlefield. But the worst of the fight waited across an adjacent crest; when the troops mounted the summit, "good God how the Balls flew."[11] John McMahon, a Civil War infantryman, fought nearly ninety years later, but battle left similarly strong impressions on the New York volunteer. The day after one 1864 action against Confederates, McMahon recorded memories of "the hard fight" in his diary. He and his unit had charged up and down two hills, exposed to rebel fire every step of the way, and eventually formed their battle line beneath a third set of Confederate breastworks. There, unprotected, they weathered a brutal crossfire that drove the Union line ahead of them backward in retreat. Combat, McMahon wrote, had thrown "a perfect hail of bullets" at the New York troops.[12] Those scenes from battlefields in the eighteenth and nineteenth centuries were not unlike one that visited an American soldier during World War II. Ordered to assault an entrenched German position, he and his comrades "charged headlong over the hedgerow into heavy enemy fire across grazed-over pastureland toward the next hedgerow," where the withdrawing enemy troops "cut us to ribbons as we ran over the open ground, charging after them. At least six enemy machine guns had us in a cross fire, and a mix of 81mm mortar, flat-trajectory 88mm cannon, high-angle 75mm howitzer fire exploded in our midst, filling the air with searing shards of shrapnel that tore through flesh and bone."[13] Another World War II infantryman described the danger of combat even more tersely: "In plain and simple words it is just plain hell."[14]

The physical manifestations of that danger, the carnage and destruction of battle, left an equally prominent impression on soldiers' accounts of combat from the War of Independence to the Second World War. In pursuit of the British after the June 1775 battle at Breed's Hill, the colonial soldier Jonathan Brigham spied "numbers of dead bodies" while harassing the enemy's retreat. The "awful solemnities" of that day, he took pains to point out in his pension application, remained impressed upon his mind more than a half-century later. The lasting memories of a battle fought some fifty-seven years earlier, "scenes of carnage and death and the inconceivable grandeur of the immense volume of flames illuminating the battlefield," remained to Brigham "as vivid as if the events of yesterday." Another Continental soldier witnessed a similarly gruesome sight passing over ground where American artillery had caught a group of enemy soldiers in a withering fire; the "dead bodies lay thicker and closer together for a space than I ever beheld sheaves of wheat lying in a field over which the reapers had just passed."[15] Such troubled recollections would have struck successors in subsequent wars as disturbingly familiar. Reuben Kelly, a Wisconsin volunteer, described the horrific conditions at the 1863 Battle of Chancellorsville as Southern soldiers attempted to assault the Union batteries: "When the rebels would get up pretty close they would open with grape and canister on them mowing them down like grass." Sparks from exploding shells ignited the brush in surrounding woods, transforming the battlefield into a literal inferno: "When our men went over to get the wounded the woods was still afire they caught from the shells bursting in the dry leaves and bodies was found all burnt up where the wounded had craweled off to get out of the way and lazed so long that they could not help themselves and a great many of the dead was burnt up." Kelly observed that his service in the army had desensitized him to such sights, noting that "when I was at home that would look hard to me but here nothing looks hard to me."[16] A soldier in the opposing army, Confederate colonel J. J. Scales, confronted similar horrors on another Civil War battlefield. Struggling to cross 500 yards of open ground in an assault on Federal artillery batteries, he watched as "men fell around on every side like autumn leaves" and noted that "every foot of soil over which we passed seemed dyed with the life blood of some one." Fully half the men in his regiment fell in the charge.[17] Similarly, a GI who landed at Omaha Beach eighty years later remembered how the sand there appeared "covered with bodies—men with no legs, no arms—God, it was awful. It was absolutely terrible."[18] The blunt conclusion of another World War II soldier, that the "human body is hard put to compete with what is devised to destroy," proved equally true and apparent to the foot soldiers of the War of Independence as well as the Civil War.[19]

Among the most obvious and salient themes in soldiers' recollections of battle is the unavoidable presence of danger and destruction. But this is by no means the only continuity in accounts of battle from different eras. Another prominent commonality, the compelling and often overpowering sensations of fear when faced with the danger and gore of battle, will be discussed in depth in the following chapter. Soldiers' attempts to put the chaos and confusion of combat into words share some specific attributes as well. The presence of weather imagery, likening a fusillade of bullets to a hail, storm, or squall, recurs regularly, as do references to battle as hell incarnate. Common subconscious actions in battle appear repeatedly across the centuries, particularly some counterintuitive ones: dropping a shoulder and pulling up a coat while trudging into enemy bullets, for example, as if the barrage of lead could be warded off as easily as raindrops in a storm. The deprivation and discomfort of the infantry constitutes another common theme, though not one that fell exclusively on frontline combat soldiers: support troops away from the firing line complained equally, if not more, about the conditions military service imposed on them. Soldiers of different eras also struggled with the necessity of killing enemy soldiers (though to widely varying degrees) and, when possible, often expressed a preference to do that killing at a distance, where evidence of the effects of their actions was less apparent.[20]

Other similarities emerge from soldiers' accounts. The overwhelming confusion of being under fire in combat—whether it was British redcoats, smoothbore 12-pound cannon, or German fighter aircraft doing the shooting—provides another theme common to soldiers' experience of battle in every era.[21] The very ineffability of battle left many soldiers unprepared to face its realities, a realization that dawned on soldiers once they had finally "seen the elephant," as Civil War soldiers referred to the first experience of combat. Imagining what battle would be like based on previous generations' descriptions simply could not convey the full magnitude of the horror that awaited soldiers when they finally stepped into battle. Horace Smith, a Civil War volunteer who fought at Chancellorsville in 1863, discovered that "I never before had any correct idea of a battle. The bullets fell like hail stones."[22] As an unnamed Confederate soldier wrote his wife a week after the Battle of Shiloh, the reality of combat was not what it was "cract up to be."[23] A veteran of Gettysburg described the battle in a letter to his wife as "awful beyond Discription." Words could not contain the horror: "I cannot discribe it with my pencil."[24] As World War II correspondent Ernie Pyle put it, being at war was a life "inconceivable to anyone who hasn't experienced it."[25]

For soldiers unprepared for the shocking realities, the carnage of the battlefield frequently overwhelmed their attempts to make sense of it. Soldiers of each era reported some kind of sensory overload in battle. Whether at Saratoga or the Wilderness or Omaha Beach, combat threw an incredible number of sights, sounds, smells, and sensations at infantry soldiers, most of them deeply disturbing. Battle immersed participants in a cauldron of powerful and terrifying stimuli, from the deafening boom of cannon and the crack of small arms to the smoke left by exploded gunpowder, to the fountains of dirt from shell impacts, and the distinct odor of black powder or cordite. Sensory data flowed in such torrents that it overwhelmed nearly every soldier in each of these wars at one point or another. There was simply too much going on at once for the individual combatant to process, one reason that participants' accounts of battle often share a disjointed quality. Smoke (particularly the dense, heavy variety produced by exploding gunpowder in these three wars) and terrain obscured large parts of the action. The noise of combat, and the deafness common in soldiers who endured long stretches in melees with firearms exploding near their eardrums, made it difficult for participants to determine precisely what was unfolding during the battle.

In the aggregate, the commonalities in soldiers' experiences are unquestionably powerful. Indeed, many of the most bracing images of battle are, on the surface, strikingly persistent across time. The description of a dirt- and powder-streaked soldier, exhausted, trembling, and vacant-eyed, might just as easily come from the line at Yorktown as the Bloody Angle at Gettysburg or the fighting in the Ardennes. Given the obvious similarities, a veteran might fairly conclude that the uniform was, indeed, the biggest feature distinguishing infantry soldiers of different wars. But to suggest that the experiences were therefore essentially the same oversimplifies the reality dramatically. Just as no historian would suggest that every manifestation of a complex institution like slavery was fundamentally alike simply because each revolved around themes of bondage, labor, and dehumanization, overemphasizing the continuities of a supremely complex phenomenon like battle to the exclusion of the differences obscures many important truths about the changing nature of the experience. Those differences play an important role in understanding how the experience of battle changed over time, since many of those changes constituted significant evolutions in the way soldiers of different eras encountered infantry combat. Often the evolutions go undetected, both by soldiers and scholars, because it is difficult to spot and make sense of the changes in isolation. Examining one battle or one war rarely reveals the multifaceted ways in which infantry combat changed over the decades. Juxtaposing different battle experiences—analyzing what the individual soldier saw, heard, suffered, and did—helps illuminate both the nature and the significance of these changes in the experience of battle.

The remainder of this chapter describes salient features of infantry combat experienced by soldiers during three wars. It introduces those qualities by briefly sketching details from one battle during each war: the Battle of Cowpens during the War of Independence, the Battle of Shiloh during the Civil War, and the Battle of Hürtgen Forest during the Second World War. Snapshots rather than exhaustive case studies, the three sketches together offer an introduction to some of the broad features that characterized ground combat in each war and a point of departure for the discussion of continuity and change that constitutes the rest of the book. The individual episodes are not intended to serve as representative of combat in a particular war; given the vast variations in combat over the course of each conflict, there is hardly such a thing as a typical battle. Indeed, each of these battles is in some ways exceptional. Cowpens stands as a model of tactical brilliance, one of the few successful uses of militia troops during the War of Independence and a dramatically lopsided American victory. Shiloh pitted armies of largely green recruits against each other, and it remains one of the most destructive battles of the Civil War: the two-day bloodbath shattered contemporaries' notions that combat would be a glorious adventure and destroyed the hopeful belief that the war would be brief and relatively bloodless. Nor can the miserable months-long struggle in the Hürtgen Forest in late 1944 be considered typical of ground combat in the Second World War. That conflict spanned continents and involved soldiers fighting in amphibious invasions, massed tank battles, airborne assaults, and thousands and thousands of smaller actions in vastly different conditions, including the heat of the African desert, the tropics of Southeast Asia, and the brutal winters of northeast Europe and Eurasia. Stretching over four months, the battle in the Hürtgen remains the longest battle the U.S. Army has ever fought. Thus these brief snapshots are not meant to serve as exhaustive descriptions; other historians have already told the stories of the individual battles brilliantly. Each of the individual scenes does, however, offer a window into the basic nature of infantry combat in each war. Placed side-by-side, these scenes provide a way to assess common threads as well as changes and evolutions in what battle required of soldiers during each era.

The 1781 battle at Cowpens took place in the sixth year of the War of Independence on a plot of land used by South Carolina farmers to graze their cattle. Some three thousand troops took part in the battle: roughly 1,900 patriots (300 Continental Army regulars, supported by militia troops) met 1,150 British soldiers on a cleared field that measured about a half-mile square. After the war, historians would judge the fight as both a masterpiece of tactical generalship, with American commander Daniel Morgan's plan to entrap his British opponent a brilliant success, and a turning point in the war's southern theater, reversing a string of British successes in the south and helping to prepare the way for the final patriot success at Yorktown the following October.[26] Few of the rank-and-file soldiers who constituted Morgan's main body on January 17, 1781, however, could have had much sense of the battle's larger strategic or historic significance. To those soldiers, the battle was terrifying, desperate, and brief: a struggle to contain their natural urges to run away from the danger long enough to bring their arms to bear against an oncoming body of British soldiers.

The arms the soldiers carried into battle helped define the way combat unfolded. Most of the Continental soldiers at Cowpens carried smoothbore flintlock muskets, a designation that referred to two distinctive features that identified the weapons. "Smoothbore" indicated that the inside of the barrel was smooth; "flintlock" referred to the firing mechanism. The flintlock solved one of the fundamental problems of handheld firearms, the need to generate sparks to ignite the gunpowder that exploded and propelled the projectile down the barrel: in the flintlock musket, that problem was addressed by means of a spring-loaded cock that held a piece of flint. Released by the trigger and driven forward, a steel hammer struck the flint, and the sharp collision created sparks that fell into a priming charge of gunpowder in a reservoir called the pan. That explosion in turn ignited the main charge that launched the projectile, a lead ball nearly three-quarters of an inch in diameter. In 1781, soldiers on both sides carried their ammunition in simple cartridges, which held a measure of black powder and a projectile wrapped in a twist of paper. To load the musket, the soldier put some powder into the pan and tipped the rest into the barrel. The soldier then jammed the ball down the barrel with a ramrod.[27] The maximum range of the weapon was about eighty yards; realistically, a smoothbore was effective to about fifty yards. A number of patriot sharpshooters, deployed in Morgan's first line at Cowpens, carried more accurate but more cumbersome long rifles. Unlike their smoothbore cousins, those arms featured a long, spiral groove carved into the barrel: this was the characteristic "rifling" that gave the weapon its name. The rifling imparted spin to the projectile as it traveled down the length of the barrel, rotation that steadied the bullet in flight and increased its range significantly. In order for the rifled groove to fulfill its function, however, the bullet had to fit extremely tightly in the barrel; a loose fit prevented the bullet from accepting the twist the groove supplied as the projectile made its way toward the muzzle. Because the shoulder arms of the late eighteenth century loaded at the same point the bullet exited the weapon, the crucial rifling forced the soldier to jam a tightly fitting projectile down the length of the barrel, often more than four feet. This was an arduous task (particularly trying in the heat of battle, as enemy projectiles whizzed past) that usually required resting the bullet on a greased swatch of cloth and the use of a ramrod. Preparing a rifle for firing forced soldiers to stand, exposed and vulnerable to enemy fire, and the residue left by the black powder propellant as it exploded fouled the barrel and required frequent cleaning. More accurate than its smoothbore cousin, the rifle was slower to load and more demanding to maintain. Like many contemporary commanders, Morgan deployed his riflemen in support roles and as sharpshooters. Musketmen armed with the quicker-firing but less accurate smoothbores composed the main body of troops.

The limitations of eighteenth-century military technologies helped dictate the way soldiers fought. In the mid-eighteenth century, an experienced soldier could employ a smoothbore musket to fire up to four shots in a minute. Loaded with buck and ball (a combination of one large projectile and three smaller ones, a mixture ordered by Commanding General George Washington after 1777), a sixty-man Continental company could deliver 240 projectiles at once if all its members fired simultaneously. That simultaneous eruption, known as a "volley," aimed to compensate for the smoothbore musket's limited accuracy. Because the interior of the barrel was smooth, the projectile it fired exited the muzzle without any spin; like a baseball pitcher's knuckle-ball, bullets from smoothbores fluttered and danced, buffeted by the wind. Combined with the eighteenth-century musketman's generally poor marksmanship (given the inherent limitations of the smoothbore itself, armies of the late eighteenth century usually placed little premium on target practice), the unsteady projectiles limited the smoothbore's accuracy dramatically. Unleashing every gun in the company in a single volley increased the chances that at least some of the bullets would strike a target.

Troops fought from formations designed to facilitate fire by volley. Soldiers in the late eighteenth century deployed in long lines, standing elbow-to-elbow with their comrades. The long dressed lines had a number of important advantages on the battlefield; in particular, the formations allowed a company of infantrymen to bring the maximum number of guns to bear against an enemy at once. Gathered into similar massed formations, the body of enemy troops made an appealing target even for the relatively inaccurate smoothbore musket—though of course the attacking foot soldiers understood that their own line appeared equally inviting to their opponents.

On the surface, the practice of deploying troops in large, exposed formations often seems counterintuitive. The modern imagination boggles at the image of troops standing so conspicuously in the open, easily spotted by their enemies and vulnerable to their fire. It would seem far more sensible for soldiers to hide behind rocks, trees, and fences, allowing them to snipe at the enemy from a place of relative safety, rather than to march out into the open in neat lines and bright uniforms. But the limitations of eighteenth-century technology demanded such linear tactics: commanders massed their men and maneuvered them into close quarters, relying on numbers to provide the mass, of fire or of bayonets, that would break the opposing line. Deploying in massed formations conferred other benefits. The close quarters of those formations meant that the individual soldier in line could spot his comrades up and down the line with a quick glance to either side, and could feel the support of his nearest neighbors next to him. Their tangible presence offered a physical reassurance during battle as well as a concrete obstacle to flight: extrication from the midst of an infantry company standing shoulder-to-shoulder in a double line was no easy feat. Deploying musketmen in blocks also offered officers and noncommissioned officers (NCOs) some ability to monitor and control their soldiers once battle began: troops grouped shoulder-to-shoulder formed a line in which shouted commands were audible, and just as the individual foot soldier could glance up and down the line for the reassuring sight of his comrades in support, gathering troops into masses allowed the officers and NCOs leading them to monitor their behavior as the battle unfolded. A soldier who found his nerve wavering as the British approached that January morning in 1781 and who wished to defect from the line understood that his actions would be both conspicuous to his neighbors and likely observed by his officers and NCOs, who were empowered with a variety of coercive techniques to counteract cowardice in the face of the enemy. The close quarters and tightly packed formations thus made running from battle a more difficult proposition in several ways.

Once it began, battle in the late eighteenth century immersed participants in a violent frenzy of sights and sounds. Given the limited size of the battlefield itself, Continental soldiers arrayed in their firing lines at Cowpens could track the approach of their British opponents, marching lockstep to a drumbeat and clad in their distinctive red coats. Officers stood near the Continental formations, urging their charges to hold their fire: the limited range of their smoothbore muskets made restraint critical, since loosing a volley too early dissipated its punch and left the unit particularly vulnerable to return fire as it reloaded. As the enemy approached within range of the line, officers led their troops through the series of commands to load and prepare their weapons. Those actions occurred in concert, with each soldier performing identical manipulations at the same moment. Officers coordinated the actions with a series of commands: "Prime and load!" followed by "Shoulder," "Make ready," and "Take aim!"[28] Nervous soldiers naturally wanted to fire early, as the imposing formation bore down on them; their officers commanded them to hold. Thomas Young, an American soldier at the battle, recalled that "Every officer was crying don't fire! for It was a hard matter for us to keep from it."[29] Up and down the line, the individual soldier could see muskets brought to shoulders in a synchronized motion; then, a pause. Finally came the shouted command: *Fire.* Bursts of thick, white smoke appeared up and down the firing line, accompanied by the cracks—not quite simultaneous, but close—of sixty guns erupting at once. The enemy line was close enough that a keen-eyed soldier peering through the heavy smoke could distinguish some of the effects of the bullets as they found their targets: wounded soldiers staggering and crumpling to the ground, the opposing line made more ragged in its movement, fountains of dirt kicked up in front of the enemy formation by bullets aimed too low.

On the battlefields of the late eighteenth century, the opposing lines usually were close enough that a Continental could see and hear an enemy formation preparing to unleash its own volley in response. That constituted as trying a moment as any in battle, since the individual soldier in the ranks could do nothing to dodge or avoid the inevitable barrage of lead. Maintaining the integrity of the firing line, without which the volley lost its effectiveness and the temptations to flee could prove overwhelming, demanded that individual soldiers remain in their ranks, upright and exposed, and absorb the enemy fusillade—in the words of one soldier of the mid-eighteenth century, "to stand still to be shot at."[30] Continental troops at Cowpens were close enough to hear an opposing leader's commands to his own troops. Since fewer than forty yards separated the two lines, patriot foot soldiers in the ranks might hear an enemy officer direct his troops through their own preparations to fire. At the shouted order, puffs of white smoke appeared from enemy guns a split second ahead of their reports. Then, a terrifying moment later, there was the noise of the volley arriving, an instant that generations of soldiers have struggled to describe. Analogies from nature and onomatopoeia abound in their descriptions: the buzz of a swarm of angry hornets, the hiss of missed shots zipping past overhead, the *thwip-thwip* of bullets striking their intended targets. A unit's first volley was the best-orchestrated of any firefight. After the initial, near-simultaneous eruption from the line, fire along the front became more uneven as casualties and the natural differences in various soldiers' rates of loading and reloading made precise coordination more difficult.

Exchanging fire from linear formations exposed troops to a wash of sensory stimuli. Battle leveled a variety of sounds at soldiers: the crack and *pop-pop-pop* of gunfire, shouted commands from their officers, the grunts of exertion from neighboring troops as they frantically loaded and discharged their weapons, and—most distressing—the cries of soldiers wounded in the exchanges. A barrage of sights confronted soldiers during the engagement, as well: most of their comrades were visible, as were most of the enemy troops. Officers and NCOs stood near their charges, dispensing encouragement, instructions, and threats; the patriot commander Daniel Morgan was conspicuous as well, urging his troops to stand fast and maintain their fire. Clouds of heavy white smoke, the byproduct of the black powder that fueled the smoothbore volleys, cloaked both lines, but through them soldiers might spot the effects of their fire: the bodies of wounded soldiers felled after each exchange. As the battle mounted, the barrage of stimuli rose to a furious and terrifying crescendo from which soldiers had little respite.

Fortunately for those troops, the firefight at Cowpens was fairly brief. Using the average time it took a soldier to reload and fire and evidence of the number of volleys the Continentals managed, one historian of the battle has estimated that the main exchange of fire lasted about five minutes. The entire battle itself was no lengthy affair: from first shot to last, it consumed just over half an hour.[31] For the individual soldier, that half-hour demanded both the physical effort of loading and discharging the heavy muzzleloader over and over, as well as the extraordinary psychological effort necessary to stand exposed and vulnerable to enemy fire as a series of volleys tore through the formation.

The close confines and limited area of the battlefield made the effects of the fight clearly visible to many of the survivors. The battle was a resounding victory for the Americans: Morgan's grasp of the limitations of his militia troops and how much hardship they could withstand helped create a battle plan that killed 110 British soldiers and captured 700 more. The American troops suffered only 12 dead and 60 wounded. For the lucky survivors among the patriot forces, the ordeal of the battle was over. But like the troops who emerged unscathed from other battles during the War of Independence, the field offered ample evidence of the fate of the less fortunate. The bodies of the battle's victims lay on the field, many with grisly wounds that testified to the extreme violence of even this comparatively brief encounter. The apparent magnitude of the victory left the American troops jubilant, but it is difficult to imagine that many of the patriot soldiers who left the field that day did not harbor the hope that Cowpens would be their last plunge into combat.

Some eighty years after the battle at Cowpens, the sun rose on another group of soldiers preparing to face battle, as more than 100,000 troops collided in western Tennessee near a small church called Shiloh. What followed was a two-day bloodletting on the banks of the Tennessee River as two Union armies totaling nearly 70,000 men clashed with some 45,000 rebel soldiers. While subsequent Civil War battles would eclipse the casualties at Shiloh, by the time the sun set on April 7, 1862, the second day of the fight, the clash along the river had extracted a staggering toll on both sides. More American soldiers died in forty-eight hours on the Shiloh battlefield than had died in every battle fought during the War of Independence, the War of 1812, and the Mexican War. Nearly 3,400 soldiers perished in the fighting, and a staggering 16,000 suffered wounds, many of them gruesome.[32]

Just as the individual soldier at Cowpens could hardly appreciate the significance of the entire engagement, no single participant at Shiloh could grasp the enormity of what unfolded all around him during the battle, much less its larger tactical or strategic significance. The two-day battle constituted a number of separate and connected smaller actions that sprawled across more than eight square miles of terrain (a few of these actions, like the struggle on a patch of ground that became known as the "Hornet's Nest" on the first day, earned their own notoriety for the ferocity of the fighting that took place there). Given its size, no individual participant could see more than a small part of the battlefield, and the deadly seriousness of combat tended to focus individual soldiers' concerns dramatically. As one Civil War veteran put it, "The extent of the battle seen by the common soldier is that only which comes within range of the raised sights of his musket."[33] But even though no single soldier could see more than a small piece of the entire battle, the localized melees that made up the 1862 battle bore enough resemblance to the fighting at Cowpens in 1781 that a veteran of that earlier battle would have found the character of combat at Shiloh recognizable in its broadest features. The weapons and tactics, while not identical to those of the War of Independence, were nonetheless similar in many ways to eighteenth-century warfare.

One obvious difference was simply the scale of the fighting itself. The battle took place on a tract of land, much of it cleared, that lay along a crook in the Tennessee River. The battlefield at Shiloh occupied some thirty times the geographic space as the action at Cowpens, and all the patriot soldiers present in that 1781 battle would have filled only a handful of the more than a hundred Union infantry regiments that fought at Shiloh in 1862. But while a veteran of the Continental Army surely would have expressed amazement at the sight of more than 100,000 soldiers locked in the desperate contest at Shiloh (the entire Continental Army never numbered more than 30,000), the tactics in use in 1862 bore a strong resemblance to those in evidence on the battlefields of the War of Independence.[34] As at Cowpens, the military technologies of the day, and their limitations, exerted enormous influence over the way soldiers deployed and fought.

Some of the Union and Confederate soldiers who clashed at Shiloh carried smoothbore muskets similar in many elements of their basic design to the weapons employed by infantrymen at Cowpens. But many units carried newer rifled muskets into the battle; by war's end, the model 1861 Springfield rifle had become the defining weapon of the conflict. It differed from its smoothbore antecedents in several important respects. Like the long rifles employed by sharpshooters at Cowpens, the Springfield rifle barrels had a long, spiral groove carved along the inside. The long rifles of the Revolutionary era had to be individually smithed, at substantial expense; by the mid-nineteenth century, rifling was standard even in mass-produced infantry arms.[35] Those rifled muskets combined with another nineteenth-century innovation, the Minié ball, to increase the range of the Civil War infantry soldier's standard arm and make the process of loading and reloading faster and more reliable. The Minié ball was a conical bullet whose base expanded and flared with the explosion of the powder charge to increase its diameter in the barrel. That slightly expanded bullet could accept the rifling, solving one important problem of the late-eighteenth-century rifle. The earlier weapon required that a tightly fitting projectile be jammed forcibly down the length of the barrel; the introduction of the Minié ball made it possible to drop a bullet down the barrel of the musket and enjoy both increased accuracy of rifling and relative ease of loading of smoothbore. There was another important difference in the arms carried by soldiers at Shiloh. Rather than the flintlock system employed by Independence-era firearms to provide the critical sparks, the newer weapons featured a percussion-cap ignition system. That mechanism consisted of a cap, usually of brass or copper and filled with fulminate of mercury, that fit on a nipple in the firing mechanism. When released by the trigger, the weapon's hammer struck the cap, igniting the black powder and launching the projectile. Because it was far more reliable than the flintlock (flints wore out, and when damp did not produce sparks dependably), the percussion cap represented a substantial improvement over the earlier system. Together, machined rifling, the Minié ball, and the percussion cap resulted in infantry fire that was both more accurate and more reliable than that exchanged across the Cowpens battlefield. For the individual soldiers, however, the operation of their arms bore a strong resemblance to the manipulations their predecessors carried out during the War of Independence. Civil War soldiers still stood to reload their weapons. While it was possible in theory to reload the rifle from a prone position, it was comparatively uncommon (indeed, the most oft-used tactical manual of the Civil War era made no provision for reloading from the ground). Like their predecessors at Cowpens, most Civil War infantry soldiers experienced much of the battle standing upright, a stance that left them exposed and vulnerable to enemy fire.

The formations employed on the battlefield at Shiloh descended directly from those used at Cowpens as well. Like their eighteenth-century predecessors, soldiers in the mid-nineteenth century deployed in dressed lines. While the size of individual Civil War units dwarfed those of the War of Independence (at the war's outset, a full-strength Union infantry regiment usually comprised ten hundred-man companies), the mass of troops usually fought elbow-to-elbow with their comrades. Both North and South recruited soldiers geographically, and because regiments formed in specific towns and cities, soldiers often went into battle alongside men they knew from home. Soldiers advanced into the melee on foot, sometimes marching a half-mile or more to engage the enemy. Officers and noncommissioned officers oversaw those movements to dress the lines, attempting to keep a uniform front among their troops. The emphasis lay on steadiness rather than speed; it was more important to keep the line together than to move it quickly, even though the modern imagination recoils at the image of an infantry company crossing a fire-swept field at a slow and deliberate march rather than racing across it to limit exposure to the hail of bullets and shells. To be effective in battle, an infantry line had to be cohesive in its actions, and a formation moving too fast or over uneven terrain usually became ragged. Since irregular lines could not deliver volleys with the same impact, and threatened to dissolve entirely under the strain of enemy fire, leaders valued steady and deliberate movement over speed.

As at Cowpens, entry into battle plunged soldiers into a cauldron of overwhelming and acutely traumatic sights, sounds, and sensations. But though battle at Shiloh was larger and louder than at Cowpens, the tactics were similar. One Civil War soldier described the appearance of those linear tactics with a diagram in an 1863 letter to a family member. The sketch depicted a cloud of skirmishers deployed before several orderly, solid lines of infantry, supported by large dots representing friendly artillery batteries. In his explanation, the author noted that linear battle had a particular rhythm: "Severe fighting does not often last at any one point for a great length of time," he indicated, "but is severe for a few moments and then subsides a short time until the forces get a little rested then go at it again."[36]

Linear formations persisted into the mid-nineteenth century because they addressed some of the lingering technological limitations of the era. Mass-produced rifled muskets, Minié balls, and percussion caps made the infantryman's standard arm more accurate and reliable, but the musket was still slow to load. Linear formations massed a company's fire, as in the War of Independence; the increased range of the Springfield rifles simply meant that the opposing lines squared off at greater distances. In addition to massing fire into volleys, linear formations continued to offer a means, however crude, to direct and monitor troops. Those measures became even more important with the order-of-magnitude increase in the scale of battle from the War of Independence to the Civil War. Packing more men with more accurate weapons onto the battlefield created a staggeringly lethal space from which a soldier's natural instinct was to run. Gathering men into firing lines provided Civil War armies with a means to communicate along with methods that kept the soldiers fighting despite powerful inclinations to flee, as subsequent chapters explore.

Standing shoulder-to-shoulder with his comrades in the line, an infantryman at Shiloh witnessed many of the same sights as his predecessors in the Continental Army. Heavy white smoke left by soldiers' exploded black powder hung in the air in front of the firing lines, and as at Cowpens individual units disgorged clouds of smoke together as they fired in unison. Packed tightly into lines, troops at Shiloh were also close enough to see the effects of enemy fire on their comrades, a particularly disturbing sight since the wounds were unusually gory. The Minié ball was a large-caliber (more than a half-inch in diameter) and slow-moving projectile; the spin that steadied the ball in flight also caused the bullet to tumble when it struck the body, leaving ghastly wounds that shattered bones and chewed flesh. Packed tightly into lines, a soldier whose neighbor was hit often found the victim's brains, blood, and viscera spattered across his own jacket. Though soldiers in both armies fitted their rifles with bayonets, Civil War battles witnessed relatively few such charges. The increase in the rifles' accuracy created additional space between opposing lines; for the most part, it had become too difficult and dangerous to cross that interval, since an opposing formation could launch two or three volleys, with deadly effect, as the attackers closed. There were plenty of exceptions, of course, and a number of Civil War battles featured hand-to-hand fighting, often of a particularly barbaric nature. In the main, however, infantry combat during the war had turned into a slugging match, as opposing lines faced each other and simply blazed away. An Iowa volunteer described what an individual soldier saw from the middle of one such massed formation in 1863: enemy soldiers appeared as "a solid wall of men in gray, their muskets at their shoulders blazing into our faces and their batteries of artillery roaring as if it were the end of the world." The rebel line stood "bravely," just over a hundred yards away.

The Union troops halted; both sides "stood still, and for over an hour we loaded our guns and killed each other as fast as we could." Casualties among comrades in the lines were plainly evident: "In a moment I saw Captain Lindsey throw up his arms, spring upward and fall dead in his tracks. Corporal McCully was struck in the face by a shell. The blood covered him all over, but he kept on firing. Lieutenant Darling dropped dead, and other officers near me fell wounded."[37]

Infantrymen at Shiloh had to contend with more than just swarms of Minié balls hurled from opposing formations. American troops in the lines at Cowpens had faced enemy cannon (the British deployed two three-pound guns in South Carolina), but few veterans of that battle mentioned the impact of those pieces on the fighting.[38] The years between the two battles witnessed enormous advances in artillery technology. The number of guns deployed, their range and accuracy, and the variety of their ammunition all increased enormously. By the middle of the nineteenth century, artillery had evolved into a supremely effective antipersonnel weapon, particularly devastating on the defensive. In the close quarters of the mid-nineteenth century battlefield, artillery loaded with canister shot (a cylindrical can filled with lead balls that turned the cannon into an enormous shotgun) became a devastating defense against attacking infantry formations. In extremely dire straits, when reduced range and faster ammunition consumption were of lesser concern, the cannon could be loaded with double, or even triple, rounds of canister, geometrically increasing its ability to punch holes in an infantry line. Joshua Callaway, an officer in the 28th Alabama, encountered one such hail of fire at the Battle of Chickamauga in 1863. The weight of the canister and small arms fire that met the members of his unit as they approached within twenty yards of a Yankee line was withering, dwarfing anything he had witnessed in his seventeen months of combat experience. It was "a scene that beggars description," he wrote his wife a few days later, "and God forbid that I should ever have to witness another. I was indeed in the very midst of death."[39] Other soldiers echoed Callaway's observations; by the Civil War, the killing power of artillery awed even experienced soldiers. At Fredericksburg, Indiana volunteer David Beem witnessed the murderous effects of a concentrated storm of artillery on his unit's neatly dressed formations. "In an instant," he explained, "shot, shell, grape and canister were poured into our ranks like rain drops and I have never yet seen so many fall in so short a time." Beem and his men continued to bear down on the Confederate position, but "all this valor, all this daring heroism could not withstand such a crossfire as their forts brought upon us, and the line stopped when it come to the rebel infantry."[40] Though the battlefields of the Civil War were larger than those of the War of Independence and the range of Civil War cannon far greater, opponents were still often close enough that artillerymen could observe the devastating effect their work had on enemy troops. As an Ohio artilleryman at Shiloh wrote in his diary just after the battle, the impact of his battery's projectiles was "too sickening for the pen to describe."[41]

Part of the terror that artillery generated in infantrymen sprang from the horrific way those weapons mutilated their victims. Gruesome dismemberment underlined the acute danger these weapons represented. While the .58-caliber Minié ball left a frightful wound, canister shot mustered dozens of balls — some (depending on the caliber of the cannon) as large as an inch in diameter. An Indiana volunteer summarized the grisly killing power of artillery loaded with antipersonnel shot after the battle of Antietam in September 1862: "Death from the bullet is ghastly," he recalled, "but to see a man's brains dashed out at your side by a grape shot and another body severed by a screeching cannon ball is truly appalling."[42] The spread of shrapnel from an exploding artillery shell added to the cannon's terrifying and deadly reach, and the red-hot temperatures of the fragments rendered them even more dangerous than comparably sized musket balls in many soldiers' eyes. One soldier recorded in his diary that pieces of a burst shell traveled up to a quarter-mile with enough power to "wound very badly." The fragments represented a twofold danger: "They come with as much force as a bullet does from a gun and when they first strike are hot, so that they both wound and burn."[43] Opposing infantry and artillery were not the only enemies with which the foot soldiers at Shiloh had to contend; both sides fielded mounted cavalry as well. While the sight of opposing troops on horseback was familiar to infantry soldiers during the War of Independence, the cavalry at Shiloh rarely fought from the saddle. Cavalry encountered great difficulty operating among the infantry lines, and at Shiloh commanders employed mounted troops principally in the rear to stop stragglers — "of whom there were many," according to Union commander Ulysses S. Grant.[44]

Soldiers fighting at Shiloh could see a good deal of their immediate surroundings. The view was hardly perfect: though generally clear, the battlefield at Shiloh featured some areas of trees, and even modest dips and swells in the terrain rendered parts of it invisible even from nearby troops. Wooded areas of the field further obscured soldiers' vision.

Once the battle began, clouds of smoke from discharged firearms made it even more difficult to see. Those obstructions only increased individual soldiers' confusion amid the chaos of battle. Many of the most obvious trappings of the Civil War battlefield aimed to counter that native confusion of combat. To help friendly troops locate each other in the frenzy of combat, soldiers wore colored uniforms, and regiments organized around a flag that provided a highly visible rallying point for the hundreds of soldiers in the unit. Drumbeats coordinated soldiers' marching, and like their predecessors on the battlefields of the late eighteenth century, critical instructions arrived via shouts and bugle commands. Altogether, an infantry regiment arrayed for battle in 1862 represented a noisy and highly conspicuous affair.

Fighting in individual actions at Shiloh usually occurred along lines where the two opposing units faced each other. While the distances between lines generally grew from the eighteenth century to the nineteenth as the range and accuracy of the weapons increased (many fights during the Civil War occurred between lines standing several hundred yards apart), some Union and Confederate regiments at Shiloh faced off as close as thirty yards in places. Across that empty space two regiments exchanged their fire in coordinated blasts of Minié balls at often terrifyingly close distances. The tactical manual of the mid-nineteenth century understood the first volley to be the most critical; in addition to its considerable destructive potential, a regiment's thousand-ball volley might carry enough psychological shock value to destroy the nerve of enemy troops, leading them to flee in panic. After the first volley, firing along the line turned more ragged: the different rates at which individual soldiers loaded and reloaded their weapons, and the growing effect of casualties from enemy fire, meant that subsequent volleys could not carry the same shock effect as the first. As the battle progressed, officers on both sides labored to maintain as constant a fire as possible up and down their companies' lines, exhorting their troops with shouted commands and doing their best to combat the nervous excitement that battle inevitably created. Though the scale of the battle was far larger, steadiness among the troops was if anything more critical within the lines of a Civil War regiment than among the troops in a Continental Army unit, and officers struggled to maintain both the integrity of the lines and their soldiers' nerve.

In most cases, those officers usually had been elected by the men themselves. Few possessed any specialized or formal training; the elections were more a popularity contest than a barometer of military acumen. (And more than a few privates were dismayed to find that the friends they had nominated for commissions became different fellows with a lieutenant's stripes on their shoulders.) In combat, the most critical duty of a lieutenant or captain was to coordinate his troops' actions, timing the launch of volleys and maintaining the integrity of the line. Rather than wield muskets themselves, the officers strode the formation and urged on their troops with shouted commands. Their exhortations proved especially crucial in the midst of the chaos and confusion of combat. Leander Stillwell, an Illinois soldier at Shiloh who attempted without success to locate a target through the shrouds of smoke cloaking both lines, heard the shouts of his second lieutenant over the din of battle, crying, "Stillwell! Shoot! Shoot!" Protesting that he could not see through the smoke to spot a target, the lieutenant urgently instructed him to "shoot, shoot, anyhow."[45] And just as General Daniel Morgan had moved among his troops in 1781, flag officers were often present in the midst of their commands even as the battle raged. At Shiloh, for example, Brigadier General Benjamin Prentiss was famously conspicuous, urging his troops to maintain their stand in the Hornet's Nest on the first afternoon of the battle. On another part of the field, Grant reported in his memoir that his lieutenant William T. Sherman was close enough to the action on the first day to be struck twice by rebel bullets (once in the hand, once in the shoulder) while a third went through his hat.[46] The overall Confederate commander, Albert Sidney Johnston, ventured close enough to the fighting that a ball struck him in the leg, fatally severing an artery.[47]

The soldiers massed in lines at Shiloh, the clouds of smoke, the cacophony of gunfire, shouted commands, and the screaming wounded: all these sights and sounds would have been familiar to a veteran of the Continental Army. Survivors of Civil War battles used familiar imagery to describe the chaos of combat they had witnessed; a barrage of bullets made a sound like rain on a rooftop, or the slap of hail against a barn. Some of the sights and sounds of Civil War battle would have struck a Continental soldier as novel (the noise of massed cannon cracking timber, the garishly colored Zouave uniforms some units wore into battle, the spectacle of a thousand-man regiment marching into the fray), but many of the most salient features of the Civil War infantry battlefield were similar to the fields of the late eighteenth and early nineteenth centuries. The tests facing the soldiers at Shiloh were similar to those their predecessors had faced at Cowpens—the physical exertion of loading and firing the heavy Springfield rifle, and the enormous psychological strain of being forced to stand upright and exposed to enemy fire—but where American soldiers at Cowpens endured only minutes of that onslaught, Civil War soldiers had to maintain their efforts far longer.

If anything, combat in the Civil War had become even more ferocious within the framework of the familiar linear tactics. Crowding more men and more cannon onto the battlefield made for more destructive and more desperate clashes, and the ability to draw on vastly larger armies for reserves and reinforcements made for more protracted struggles. At Shiloh, no individual action was more ferocious or more desperate than the area of the field that came to be known as the Hornet's Nest. There, along a mile of frontage held by 4,500 Union troops, Confederate commanders leveled a savage attack intended to dislodge the determined defenders. Union general Prentiss had received orders to hold the position "at all hazards," and his command organized a steadfast defense as some 18,000 Confederate soldiers attacked in twelve separate assaults—a few coming close enough, in one Union soldier's estimation, to lay hands on the defenders' cannon. Finally, in the afternoon, the rebels massed sixty-two field guns less than a third of a mile from the Union line and hammered away at the Federal position. With each gun firing three or four shots per minute, together the five dozen guns could muster more than 180 shots every sixty seconds. The deadly barrage lasted an hour, twice as long as the entire battle at Cowpens. Finally, after enduring a mass of fire that few in the Continental Army could have conceived, Prentiss surrendered the survivors. Fewer than half his men remained.[48]

American soldiers who emerged from the half-hour of shooting at Cowpens unharmed went to bed that night knowing that they had survived the battle. Union and Confederate soldiers who lived through the day at Shiloh had no such guarantee. The fighting at Shiloh, as in most Civil War battles, lasted significantly longer than the battles of the War of Independence, and the 1862 battle was one of a growing number of multiday fights during the war. Battle raged from about 6:30 in the morning until dusk on the first day, and the troops in both armies endured an anxious and miserable night bivouacked on the field (in his memoir, Grant recounted that rain fell in torrents and that the troops "were exposed to the storm without shelter") before the Union counterattack resumed the battle at sunup the following day.[49]

The conclusion of the battle on April 7 left the survivors with horrifying and unavoidable evidence of combat's destructive power. More than 3,000 lay dead (battle deaths at Shiloh numbered more than the total number present at Cowpens), and nearly five times as many lay wounded. Those unfortunate victims often suffered staggeringly gruesome wounds. Tumbling Minié balls, red-hot shrapnel, and large-diameter canister rounds inflicted frightful damage on vulnerable bodies. Soldiers shot in the gut went septic, left to die an agonizing and untreatable death; those shot in the arm or leg frequently discovered in a hastily erected field hospital that the medicine of the mid-nineteenth century had only amputation to offer.[50] The agonized cries of the wounded and the dying pierced the air, and even those lucky enough to emerge from the fight relatively unscathed nonetheless suffered its effects. Most were exhausted, physically and emotionally; many suffered from acute dehydration. The concussive effects of the weapons left noses bloodied and eardrums shattered; some soldiers wandered back from the lines in a daze with blood streaming from their ears, and many more suffered from a temporary "battle deafness" caused by hundreds of gunshots discharged in the close quarters of the line. A host of emotions accompanied the physical effects of battle. Not surprisingly, profound relief constituted one of the most common responses; many of the survivors of the battle, green recruits at its outset, evinced horror at the sights and sounds of battle and expressed the hope that Shiloh would be the only such terror they would have to witness. Though no one knew in April 1862 that the war would rage three more years, some of the veterans of the battle were already waking to the realization that the notion of a brief, glorious struggle had become an unlikely hope indeed.

Eight decades passed between the battle at Shiloh and the battles of the Second World War—the same amount of time separating Cowpens from Shiloh. But where a veteran of Cowpens would have found the character of Civil War battle roughly familiar (if even more disturbing in its scale, duration, and gore), by the mid-twentieth century the nature of infantry combat had changed so dramatically that Union and Confederate veterans would have found it all but unrecognizable. Indeed, many American GIs who fought in the Second World War, raised on images and stories of combat from the Civil War, found the reality they encountered in the mid-twentieth century startling. One foot soldier registered surprise at the appearance of the battlefield; having pictured combat as scenes from D. W. Griffith's 1915 film *The Birth of a Nation*, with "rows of men advancing shoulder to shoulder, cannons firing, the first ranks being melted down," he was unprepared for what he found in the European Theater, where he and his unit constituted but a "lonely little string of men."[51]

Part of that surprise stemmed from differences in the magnitude of the battle. Just as Shiloh took place on a vastly larger battlefield than Cowpens, involved more troops, and lasted far longer, battles of the Second World War eclipsed those of the Civil War in scope. But it was not merely the scale that would have stunned veterans of earlier battlefields; for the individual soldier caught in its grasp, combat looked and felt immeasurably different. Battles of the War of Independence and the Civil War frequently resembled large mobs crashing into one another. Banners provided a visible point of reference around which infantrymen formed their lines, and colored uniforms distinguished friend from foe. Most importantly, units met in the open; leaders were often visible directing their units as they poured fire into enemy formations. By comparison, World War II battlefields appeared much more sparsely populated. Dressed lines, advancing by ranks in the open as they had on the battlefields of the War of Independence and the Civil War, almost never appeared on the fields of the Second World War. On the unusual occasions when they did, the results usually proved catastrophic.

One cluster of GIs during the Second World War learned precisely how ill-suited eighteenth- and nineteenth-century tactics were to twentieth-century battle when, improbably, a German unit deployed in the neat lines of earlier battlefields before an American unit in 1944. Fifty years later, the attack remained one GI's most vivid recollection of the entire war. Four or five hundred yards from his position, 900 German soldiers marched forward steadily in four lines, with twenty-yard intervals separating the ranks. Officers "walked just in front of each wave to keep it in perfect alignment," exactly as they had on the battlefields of the eighteenth and nineteenth centuries. To the American witnesses, the recklessness of the attack was "inconceivable." Though the Germans were "numerically strong, well disciplined, backed by plenty of artillery, brave and willing to die," they were nonetheless marching against troops armed with machine guns and automatic weapons, men who could call on four dozen artillery pieces from a supporting division and a battalion of indirect-fire mortars. The result of the German attack, predictable to anyone familiar with the killing power of World War II arms, was both quick and severe: GIs "began to pour a devastating mortar, rifle and machine-gun fire into their ranks, which they reformed without pause over the bodies of their dead and wounded." As the mass of German troops moved inexorably forward, the Americans continued to direct their fire into it; one GI's machine gun grew so hot from the continuous firing that he could no longer hold it. Aided by their artillery support, the American soldiers "fired with deadly precision at the iron-nerved, obedient men walking across the field to the certain death of sheep following the leader to the slaughter pen." The intensity of the fire created a "hell of shrapnel" in which it seemed that "not even a fly or cockroach could survive." Within ten minutes nearly every German soldier in the attack had been killed by the barrage of American fire.[52]

The abject futility of that attack, the product of eighteenth-century tactics disastrously misapplied on a twentieth-century battlefield, underscored the significance of the changes that had occurred in the infantry soldier's environment. Most obvious to the individual soldier was the fantastic danger of combat: the sheer number of rounds that twentieth-century automatic weapons were capable of firing created what one World War I trooper referred to vividly as the "storm of steel."[53] Over time, the advent of that storm required commanders to revise their tactical ideas dramatically. Armies of earlier eras had relied upon close-order linear tactics to maintain unit integrity and discipline on the battlefield and to deliver the mass, either of men or fire, to break the opposing force. While armies of the mid-nineteenth century often sustained enormously heavy casualties while exposed to enemy fire in their dressed lines, they were nonetheless able to maneuver, deliver return fire, and charge with some degree of effectiveness. As the GIs pouring fire into the German ranks observed, the lethality of automatic fire and artillery rendered those tactics suicidal: no massed body of troops in the open could hope to withstand the murderous fire of automatic weapons long enough to muster an effective attack.

Because the storm of steel was too dangerous for massed bodies of troops to weather in the open, infantry soldiers could no longer fight from those formations. Instead, troops spread out from their lines so that no single machine-gun burst or artillery blast could annihilate them all. Those dispersed tactics (adapted from methods first explored during the Boer War in the late nineteenth century and employed with great effect by German storm trooper units at the end of the First World War) allowed soldiers to mitigate the fearsome destructive power of automatic weapons to some degree. Infantry soldiers operating in the dispersed system of tactics used a combination of techniques to retake the initiative. Cover (interposing an obstacle between the defender and thereby denying a clear line of sight) and concealment (the use of camouflage to deny the defender an obvious target) helped neutralize the defender's immense advantage in firepower. Rather than attack in massed and slow-moving human waves, soldiers in the dispersed system utilized coordinated move-and-fire techniques: working in small teams, one element of the attack provided suppressive fire, forcing defenders to momentarily duck their heads, while another element used the respite to sprint for cover. Once there, the elements exchanged roles, the latter providing suppressive fire while the former raced for a new refuge. Working together, coordinated teams could thus keep enough pressure on opponents to offset the higher rates of fire those defenders enjoyed due to their automatic weapons.

The overall effects of this revolution in tactics were immense, and successive chapters explore some of the most noticeable from the point of view of the individual soldier. Nowhere were the effects more visibly obvious, however, than in the dramatically altered appearance of the battlefield. To the lone infantryman in the winter of 1944, battle looked and felt vastly different than it had to soldiers in 1781 and 1862. Rather than an open field filled with neatly dressed lines, brightly colored banners, and leaders on horseback, the individual infantryman saw much less overt activity on the battlefields of the Second World War. Both friendly and enemy soldiers spread out and sought cover to reduce their exposure to deadly fire, leaving the individual soldier much more alone. The battlefield that had appeared in the eighteenth and nineteenth centuries as a clash of two armed mobs in the open had become by the mid-twentieth century a seemingly desolate space. Dispersed and camouflaged soldiers sheltering from enemy fire produced the twentieth century's distinctive "empty battlefield." World War II soldiers who expected the massed lines of earlier battlefields found the emptiness disconcerting. After his first experiences in battle, World War II soldier John Babcock admitted, "What I thought this war would be is all turned around." The impression he had received from books and movies included "a beat-up field, peppered with shell holes, and guarded on both sides by big, old trenches," and "some hazy idea we'd skootch down in one row of them until Captain Goodspeed blew the whistle" to go "over the top." Those notions left him completely unprepared for the reality of combat on the empty battlefield: "When the real thing came, I thought we were still just jockeying around that morning when we started through the new snow. . . . I couldn't believe that this was the real front until I started to get pasted by MGs and artillery."[54] "Forget the panorama" of battle as pictured by Hollywood, wrote another GI. "Battles are fought by small groups of desperate men against other small groups equally desperate," who were "infrequently able to see their comrades further along, or even the enemy across the way."[55]

The American soldiers who fought in the battle in the German Hürtgen Forest during the winter of 1944–1945 experienced the trials of the dispersed battlefield firsthand.[56] That battle was far larger (sprawling across more than fifty square miles) than battles of the War of Independence and the Civil War, and it lasted far longer: American units fought for weeks rather than days at a time. The 200,000 soldiers engaged there understood that they constituted only a tiny piece of a gigantic global war, raging simultaneously on the German border, on the Eastern front, on the Pacific islands, and in Southeast Asia as well as in the skies over Europe and the waters of the Atlantic and Pacific oceans. For troops who had witnessed the D-Day landing the previous June with hope that the war would be over by Christmas, the demoralizing realization that the German army still possessed plenty of fight had sunk in. Allied units would have to slog through Europe and into the German heartland, wresting territory from a defender determined to extract a high price for it.

What the slog of infantry combat demanded of those American soldiers differed in important ways from what the battles at Cowpens and Shiloh had demanded of their predecessors. Many of the changes originated with the weapons technologies of the mid-twentieth century; as during the War of Independence and the Civil War, the military technology of the day and its limitations exerted a profound influence on the way soldiers experienced battle. Among the most pronounced changes from those earlier conflicts was the introduction of automatic and semiautomatic weaponry. Those weapons forced a new bullet into the breech of the gun as soon as one was fired, eliminating the lengthy and physically demanding reloading that muzzleloaders had required of soldiers in the eighteenth and nineteenth centuries. A majority of American GIs in the Hürtgen carried the M1 Garand rifle, which accepted an eight-shot clip. The semiautomatic action allowed a soldier to fire all eight shots as quickly as he could pull the trigger, without standing upright and forcing each individual bullet down the length of the barrel. The result was an order-of-magnitude increase in the amount of lead an individual soldier could put in the air—and because the bullet traveled at such a high speed, the Garand boasted both a longer range and more accuracy than earlier shoulder arms. A twelve-man squad of GIs in 1944 could thus fire as many shots in a minute as an entire Civil War company. And unlike their predecessors, infantry regiments in the Second World War carried a far greater variety of weapons. Though the majority bore the M1 rifle, other members of the squad, platoon, and company carried pieces of mortars, portable machine guns, and other heavy weapons that they could assemble and utilize flexibly, along with explosive grenades. Some soldiers carried a completely different weapons system; one man in each infantry squad, for example, carried the Browning Automatic Rifle, a heavier and more imposing weapon than the standard M1.

Warfare in the Hürtgen battle exhibited other immediately apparent differences from combat in the War of Independence and the Civil War. For individual soldiers, it was both terrifying and ongoing, with combat lasting for weeks rather than minutes or days. Soldiers also found the pace of battle uneven; periods of tedium alternated with periods of terror, and the shift from boredom to panic was often unpredictable. Unlike the cleared fields where battles usually took place in the eighteenth and nineteenth centuries, the terrain in the Hürtgenwald (as in many World War II battles) boasted fearsome obstacles: thick forest, steep hills, deep mud. Other changes in the nature of ground battle were more substantial. The necessity of utilizing dispersed tactics in the face of the awesome firepower placed a new premium on stealth, for the simple reason that undetected soldiers offered few targets. Indeed, the success of the tactics of dispersion depended on soldiers hiding their location from enemy gunners. That represented a profound departure from earlier foot soldiers' experience of battle. Infantry formations in the War of Independence and the Civil War massed in the open, deriving little benefit from being hidden; indeed, insofar as stealth threatened the integrity of the formation, it was a serious liability to the unit's function. The drums, bugles, uniforms, and flags of the linear battlefield were designed to be as conspicuous as possible: Infantry soldiers in those units required clear signals to tell them where to go and what to do, and those cues also provided valuable reassurance to the men to keep fighting. But the conspicuousness that was crucial on the linear battlefield proved suicidal on the dispersed battlefield. Dispersed tactics demanded that soldiers attempt to disguise their presence and their movements as much as possible; soldiers in the open were simply too vulnerable to survive for long.[57]

The new emphasis on stealth also required the soldiers fighting in the Hürtgen Forest to consider their actions, and what those actions might reveal about their location, far more carefully. By rendering the old massed regimental attacks of the linear system obsolete, dispersed tactics forced infantry units to break their assaults into smaller units: platoons, squads, and pairs. Those smaller elements could exploit the dead ground in the landscape—a shell hole, a small ridge—that of-fered small groups of soldiers shelter from enemy fire but which could not hide larger formations. In that sense, each offensive action became a miniature tactical problem. Where combat required soldiers of the War of Independence and the Civil War to muster the willpower to stand exposed to waves of enemy bullets, fighting within the dispersed system demanded that small groups of soldiers engage in a sophisti-cated intellectual exercise: deduce where hidden defenders sat, locate dead ground, find spots from which to defilade an enemy position. Those demands forced individuals and groups of soldiers to improvise on the spot and to make a number of complex, independent decisions. The new technologies and terrains that ground combat encompassed in the mid-twentieth century produced a host of tactical problems for infantry units to solve. GIs in the European Theater of Operations (ETO), for example, had to improvise effective techniques for attack-ing northwest Europe's distinctive hedgerow country. Troops attack-ing German pillbox emplacements—which typically boasted eight to ten inches of steel armor covered with several feet of reinforced concrete and dirt—devised a six-step method for neutralizing those defensive fortresses using a combination of small arms and explosives through a costly process of trial and error.[58] The need to treat tactical situations as puzzles and the necessity of experimenting with ways for small groups of soldiers to attack them would have struck the rank-and-file of late-eighteenth- and mid-nineteenth-century armies as wholly alien: for them, such improvisation was not only unnecessary but potentially disastrous.

For the same reasons, move-and-fire tactics also rendered com-munications among friendly troops more critical, even as they made them more difficult. Coordinating the actions of individual groups of soldiers (orchestrating one element's dash for cover while a compan-ion element provided the crucial suppressing fire) was instrumental to the individual soldier's survival, as well as to the successful pursuit of an objective. But the traditional modes of communication on the linear battlefield—bugle calls, regimental flags—became impossible, since they promised to reveal a position to the enemy. Even shouted commands carried new danger. Soldiers fighting within the dispersed system had to employ whispers and hand signals to communicate with friendly troops without revealing information that could compromise their safety. (A raw recruit in basic training marveled at how an expe-rienced squad choreographed their movements in one demonstration: "they never made a sound and exchanged only a few hand signals," but "if a shot was taken at them, every window in the area blazed with rifle fire.")[59]

Evidence of the frightening power of enemy weapons surrounded infantry troops in the Hürtgen Forest. William Devitt, a lieutenant who fought in the Hürtgen, remembered that his first impression of the battle was the "unremitting noise of the artillery," which sounded like "a thunderstorm that went on and on without stopping."[60] Artillery of the Second World War made such an overwhelming impression in part due to advancements introduced during the First World War and refined in subsequent decades. Improvements in the guns' range, accuracy, and destructive power gradually transformed the cannon of the Civil War—already devastatingly effective against infantry, particularly in a defensive role—into the supreme killing tool of the Second World War. Taking full advantage of those improvements in artillery technology, however, demanded some parallel advances in infantry tactics. As a number of World War I battles had demonstrated so vividly, massive preparatory barrages lasting hours or days could not destroy opponents' defensive positions entirely; enemy troops took refuge from the shells in bombproof dugouts and prepared for the imminent approach of the attacking infantry. Artillerymen of the First World War gradually developed techniques that allowed them to employ their batteries more effectively: the rolling barrage that advanced steadily across no man's land toward the opposing line with the infantry following close behind, and the hurricane barrage, which eschewed lengthy periods of shelling designed to destroy the enemy position with a brief attack designed to suppress defenders long enough for the infantry to advance and engage them.[61] Those techniques, employed with devastating effectiveness by Americans and their opponents during World War II, required not just improved artillery pieces and communications technologies to connect frontline troops and forward observers with batteries in rear areas but also, as chapter 6 explores, a new kind of coordination and trust among friendly soldiers, often at great remove.

Indeed, by the Second World War, the weaponry had become so powerful that it possessed the power not only to alter the landscape but also to make parts of the landscape itself deadly. Emblematic of this evolution was the widespread use of the tree burst, in which gunners timed a high-velocity, high-explosive artillery shell to detonate high in the tree canopy underneath which enemy soldiers cowered for protection, sending thousands of deadly splinters showering down onto opposing troops. The tree line, which had once represented a safe haven to soldiers, became its own killing field, as the fusillade of natural, slivered shrapnel that accompanied a shell's explosion magnified the artillery's already massive destructive power.[62] Veteran soldiers learned that though infantrymen "instinctively seek the shelter of trees" when threatened by an attacking enemy force, just as they would during a thunderstorm, hiding in the woods during combat actually multiplied rather than alleviated the danger. To the lone infantryman, the effect of tree bursts was terrifying: "Shells hitting treetops explode high above the ground like giant shotguns, driving shell fragments down into foxholes."[63] The wooden shrapnel was every bit as deadly as its metallic cousin. World War II officer Paul Fussell described the effect of one such tree burst in northwest Europe: while lying flat between two comrades, so close their shoulders were touching, "a shell hit the tree 30 feet above us, an immense crack and bang that left me deaf for five minutes." The two comrades were killed instantly, while Fussell took shrapnel in his back and leg.[64] Repeated exposure to such attacks drove home a terrifying point: modern weapons had enough power to turn even a seemingly harmless object like a tree trunk into a deadly weapon. The increased power of twentieth-century weaponry removed many formerly safe refuges from the battlefield.

The fearsome power generated by the new small arms forced soldiers to seek shelter at every opportunity, and the necessity of employing cover and concealment on the battlefields of the mid-twentieth century meant that the perspective of soldiers in the Hürtgen Forest differed starkly from that of their predecessors on the linear battlefield. The high volume of fire produced by automatic weapons forced troops to go to ground regularly: a prone figure presented a much smaller and less inviting target to enemy gunners and proved much more difficult to hit. That simple shift in perspective altered the World War II soldier's view of his surroundings dramatically. On the ground, even small dips and swells in the terrain created patches of dead space invisible to the prone soldier; reducing an enemy's ability to spot him simultaneously reduced what the soldier himself could see. Even the comparatively smaller outline of a soldier on the ground was not always sufficient to provide refuge from the fearsome destructive power of twentieth-century weapons. To further lessen their exposure to hostile fire, troops of the Second World War dug themselves into the earth whenever possible. Digging in was not new to World War II, of course: in the latter half of the Civil War, soldiers often employed makeshift trenches and dugouts to shield themselves, and the elaborate trench networks of the First World War came to define ground combat in that conflict. By World War II, such protection had become an absolute necessity for survival. Thus a detailed flyer circulated to American soldiers in 1943 instructed them to "DIG! . . . OR DIE!" and taught them the proper techniques to dig holes that could provide shelter from small arms fire, mortar shrapnel, and "the crushing action of tanks." Increases in the volume of fire forced infantry troops to improvise their own subterranean defenses constantly as they moved forward, and the soldier's portable entrenching tool became an indispensable part of his gear. A GI described his relationship with the small shovel that hung "affectionately in a special holder in my pistol belt" in emotional terms: "I began to know, hate, and then love that little trench shovel." The implement grew worn and shiny from use, and eventually it was replaced: "There was to be another like it later, more worn and shinier. If I was engaged to the first one, I married the latter."[65] The immense volume of fire an opposing army equipped with automatic weapons and high-powered artillery could maintain combined with the necessity of digging in to create another phenomenon uncommon on the battlefields of the eighteenth and nineteenth centuries: being pinned down. Troops who sought cover or went to ground to avoid the deadly gaze of opponents' guns sometimes found that the enemy could maintain enough fire to keep them there. Without a distraction to break the incoming suppressive fire, troops could be stuck—vulnerable to mortar and artillery rounds and often helpless—indefinitely. The limited perspective afforded troops pinned down on the ground made it even more difficult to determine when the coast had cleared and invited opportunities for enemy trickery as opponents tried to lure soldiers from their holes.

In most cases, the comrades who shared those holes were not neighbors and acquaintances that the soldiers knew from their hometowns. Because their armies recruited geographically, soldiers of the War of Independence and the Civil War went into battle alongside men familiar to them from civilian life. But in part because of the nation's experiences during the Civil War (where a regiment's unlucky placement in a particularly dangerous spot on the battlefield might result in the deaths of an entire town's young men), the American army of the Second World War mixed its soldiers into units alongside troops from a variety of states and towns so that the costs of battle did not fall disproportionately on any one locale. The junior officers were no longer figures soldiers knew from their civilian lives, and the nature of the leadership those officers provided during battle in the Second World War differed dramatically from that visible on the linear battlefield.

Unlike their predecessors, the officers of a Second World War infantry unit were not selected by popular vote. Nor did the army place them in command of their units with scant preparation, as had been common during both the War of Independence and the Civil War: officers of the mid-twentieth-century American army emerged from careful screening and much more thorough preparation, some at the United States Military Academy, far more at Officer Candidate School.

In combat, though, the rigorous selection and training their lieutenants and captains had undergone proved all but invisible to the men they led. The nature of the leadership they provided was far more obvious. Here again the new character of infantry combat placed demands upon those officers that their equivalents in the War of Independence and Civil War armies would have found completely alien. Rather than simply coordinate volleys and urge their men to keep up a steady rate of fire as enemy bullets plowed into the line, officers on the dispersed battlefield had to master a wide range of skills and make a variety of split-second judgments: where to place the men, what avenues of attack to pursue against a machine-gun emplacement, the safest route to a particular set of map coordinates. And, as chapter 4 explores, the specific tools available to those leaders evolved significantly from the linear system to the dispersed system. At Cowpens and Shiloh, some soldiers had been able to spot their commanding generals not far from the firing line, taking encouragement from their presence; such appearances were exceedingly rare in the Second World War.

There were other qualitative changes in the nature of the combat experience. Unlike their predecessors, soldiers of the Second World War fought against and alongside machines. While the unforgiving terrain of the Hürtgenwald made the wide-scale use of armor difficult, the GIs mired in the fighting endured fire from some German Mark IV tanks; a few American units received support from their own light armor.[66] Mechanized combat added a variety of new sights and sounds to the hurricane that engulfed soldiers in battle: the molar-rattling noise of an unmuffled diesel engine, or the terrifying profile of a 45-ton armored behemoth (impervious to the infantryman's small-arms fire) cresting a nearby ridge. (Mobile tanks introduced new smells to the infantry battlefield as well. For the first time the odor of exhaust mingled with the scent of gunpowder in the infantryman's nostrils. Exploding tanks added the stink of flame: no smell was more revolting than the stench of bodies burning, unable to escape their armored coffins. Medics of the Second World War learned to approach burning tanks from the upwind side so as not to be overcome by the noxious smells.) And in another dramatic departure from the experiences of the Continental Army and the Civil War armies, the introduction of machines to the battlefield added a third dimension to combat. No longer confined simply to the two-dimensional surface of the terrain, the appearance of aircraft brought battle overhead. Unlike ground troops in the War of Independence and the Civil War, soldiers of the Second World War often had to look up to locate an enemy menace.

For the individual soldier, the battlefields of the Second World War—in the Hürtgen Forest and elsewhere—had expanded enormously. Battle that had once occurred along the line where two infantry units faced off in opposing formations had expanded beyond that one-dimensional line. Armies now employed defenses in depth, so that breaching the enemy line became the beginning of an attack rather than its successful conclusion. The combination of barbed wire, concrete pillboxes, reserve formations, and other industrial innovations meant that the battlefield often extended miles back from the front. The fearsome reach of mid-twentieth century artillery meant that areas even miles behind the lines were potentially lethal. The twentieth-century battlefield harbored dangers beneath the earth's surface as well as above it. Buried land mines rendered empty fields treacherous: a misstep might trigger the hidden explosive device, killing or maiming the victim. This new variety of danger degraded soldiers' morale in a different fashion than other, more obvious threats. As combat journalist Ernie Pyle pointed out, "The greatest damage was psychological—the intense watchfulness our troops had to maintain."[67]

In the aggregate, these changes transformed the experience of battle for the foot soldier profoundly. One thing that changed very little, however, was the obvious danger of battle. Captain Donald Faulkner recalled surveying "a picture of real carnage" in November 1944, as he and his company battled German resistance in the dense forest of the Hürtgen. The power of World War II military technology was in stark evidence ("Blasted trees, gaping shell holes, and the acrid smell and smoke of small arms and mortar fire completed the terrible scene"), and the human cost of the battle lay everywhere: "arms, equipment, dead and wounded, Jerrys and GIs strewn all through the woods."[68] The mangled limbs and scorched bodies recalled similar victims on earlier battlefields. If anything, the fearsome power of the twentieth century's new weaponry added a horrifying new range of injuries to the infantry battlefield: bodies burned unrecognizable by jellied gasoline or reduced to "pink mist" by high explosives. But those same technologies also introduced a new category of battle injury virtually unknown in eighteenth- and nineteenth-century combat: the "million-dollar wound," an injury serious enough to warrant a soldier's removal from combat without killing or permanently disfiguring him.[69] Such a wound was essentially unknown during the Civil War; the soft-lead, large-caliber, slow-moving Minié ball often tumbled as it struck flesh, chewing muscle and crushing bone. Many soldiers hit in the torso died instantly; many more died later from infection that nineteenth-century medicine could not treat. Those struck in the arm or leg usually faced amputation—and, not uncommonly, death from a secondary infection. In contrast, the smaller-caliber, faster-moving, jacketed ammunition of the Second World War sometimes could punch a clean, if painful, hole through a shoulder or thigh: a wound the army might deem serious enough to warrant discharge from combat but which, thanks in part to improvements in battlefield medicine, might leave few permanent ill effects worse than a slight limp. Though comparatively rare, such million-dollar wounds offered World War II GIs the tantalizing possibility of an honorable exit from battle. As one GI recalled, he and his comrades sometimes mused on the prospect appreciatively: "Occasionally, we'd talk about the 'million-dollar wound,' shot in the foot and sent home."[70]

Of course, only a tiny fraction of the combat wounds inflicted during the battle in the Hürtgen were of the million-dollar variety; the vast majority proved far more damaging. Over three months of fighting, American units suffered more than 32,000 casualties in the savage exchanges there. Much had changed in the way Faulkner and his soldiers met and fought their enemies, but his overall summary of the battlefield recalled the impressions of soldiers surveying battles fought 80 and 160 years earlier: to Faulkner, "It just looked like hell."[71]

As the snapshots of battle from Cowpens, Shiloh, and Hürtgen Forest suggest, there was certainly no shortage of common threads, even among battles separated by a century and a half. Confronted with the deeply unnerving sight of a bloated, discolored body shot through the head or with a leg torn off from an artillery blast, the scholar can hardly fault an infantry veteran for suggesting that, from the combatant's point of view, it was the uniforms that constituted the most significant change from war to war. But probing beneath the surface of the experiences reveals a host of broad changes that altered the fabric of soldiering fundamentally as technology changed the patterns of warfare on the ground. Four in particular stand out.

As technological improvements steadily increased both the volume of fire armies could generate and the range of their arms, combat for the soldier on the ground became increasingly depersonalized. As the separation between friend and foe and the importance of seeking cover and concealment grew, the enemy became increasingly distant and faceless. In one sense, the depersonalization of twentieth-century combat formed part of an ongoing progression that stretched back to ancient battlefields. Infantry battle that had once been a hand-to-hand melee between enemies at arm's-length gradually became less and less intimate. The introduction of missile weapons and then gunpowder increased the separation between combatants: lines that had met at thirty yards during the War of Independence faced off at a hundred yards during the Civil War until, by World War II, foot soldiers often fought against a faceless, and frequently unseen, enemy. A World War II GI summarized the impression of being on the receiving end of such long-range fire from a hidden position when he described a sergeant who had "simply been snuffed out by a burst of machine-gun bullets he never saw coming from a Kraut soldier who probably didn't know his bullets were striking a human target."[72] The increasingly depersonalized character of infantry combat affected the way troops responded in battle by presenting a growing number of threats to which the individual soldier could not respond directly. A Continental soldier could usually see his antagonist, and (at least in theory) could respond to the danger by returning fire. The World War II GI targeted by aircraft, armor, or long-distance artillery often found himself the target of a distant enemy with no way to answer the threat.[73] Buried antipersonnel mines represented the lethal apotheosis of this trend: a deadly weapon whose operator was absent entirely, completely removed from the victim's reach. In addition, as ground combat became more depersonalized, individuals on the ground experienced a corresponding, demoralizing sensation of insignificance, the feeling of being just a tiny piece of a vast and unsympathetic machine. Gerald Linderman's descriptions of battle in the Second World War, *The World within War*, includes the story of an unusual but terrifying way that an unlikely mistake reminded some unlucky soldiers of the massive scale of conflict in the Second World War and of the war's vast indifference to their own plight. Anticipating their participation in a large-scale offensive, a group of American soldiers waited for their support batteries to "soften up" the enemy infantry opposing them. To their horror, artillery shells began landing on their own position. Frantic phone calls indicated that the explosions were, in fact, friendly shells falling too short: one of the hundreds of artillery tubes contributing to the preattack bombardment had the wrong range. In a grotesque twist that compounded the unfortunate targets' horror, friendly commanders had no way to determine exactly which of the batteries had erred. Ending the short-shelling ordeal of this one American unit would necessitate ceasing the entire preparatory barrage, a practical impossibility that would jeopardize the entire attack. The horrified GIs thus were forced to endure a bombardment by their own artillery, powerless to move or to stop it. The incident provided a vivid illustration of the enormous scale of twentieth-century industrial war and the soldiers' own insignificance within it.

Advances in weapons technologies and the new tactics they demanded contributed to a second change in the pattern of ground soldiers' experience of combat: a pervasive sense of isolation in battle. Dispersion meant that individual soldiers grew distant and more isolated not just from their enemies but from their comrades as well. Soldiers in the Continental and Civil War armies fought in a crowd; World War II infantry armies abandoned those crowded lines for dispersed formations that helped protect the unit from the potentially devastating effects of automatic fire and exploding artillery shells. Comrades and enemies alike were spread out, camouflaged, dug in, and hidden from view. That isolation had a profoundly disturbing effect on most soldiers; even for a man who was accustomed to fire, argued the scholar and observer S. L. A. Marshall, "experiencing it when he is alone and unobserved produces shock that is indescribable."[74] One World War II paratrooper fighting at dusk voiced concerns common in the dispersed system: "It had grown dark. Suddenly I felt alone. Where were my comrades?" Alone, unsupported, and vulnerable, the trooper felt "helpless. I couldn't fight back. I was alone in a fiery maelstrom."[75] "The psychological isolation," wrote another GI, "is even more dramatic than the physical isolation." That isolation narrowed the focus of the soldiers (perceiving themselves alone, soldiers were concerned mainly "with the few square yards surrounding them") and triggered nagging anxieties. "Our mortars and MGs—where are they? And our artillery support? Has everybody gone the hell home?"[76] Combat on the linear battlefield was many things, but it was rarely lonely. French soldier and military theorist Charles Ardant du Picq anticipated that feeling of isolation in twentieth-century battle at the end of the nineteenth, noting that open-order battle made soldiers feel adrift: the soldier was "lost in the smoke, the dispersion, the confusion of battle. He seems to fight alone." That isolation aggravated the already unsettling sensation of uncertainty in combat. In the linear system, soldiers could monitor the behaviors and fates of their neighbors, but as Ardant du

Picq noted, such reliable policing proved impossible on the dispersed battlefield: "A man falls, and disappears. Who knows whether it was a bullet or the fear of advancing further that struck him!"[77] Such anxieties gnawed at the minds of soldiers in isolation in a way unfamiliar to their predecessors.

A somewhat paradoxical third evolution plagued troops on the increasingly depersonalized and isolating industrial battlefield: a sense that combat felt frighteningly specific to individual soldiers. Increasingly, infantrymen experienced the sensation that enemy bullets had singled them out personally. That was a comparatively rarer impression within the close-packed ranks of the linear battlefield: the relative scarcity of targeted fire on the battlefields of the eighteenth and nineteenth centuries, and the way in which linear tactics forced formations to stand and absorb the volleys of their opponents, reduced the impression that any particular soldier was the target. The experience of being shot at in combat evolved steadily over time. The rifled muskets of the Civil War were far more accurate than the smoothbores of the War of Independence; Continental soldiers in the ranks often aimed at nothing more specific than the enemy mass itself, while some Civil War soldiers attempted to single out a particular enemy soldier or officer. A Civil War soldier in the ranks recalled that smoke and confusion frustrated his unit's attempts to target soldiers specifically: "The Rebels in front we could not see at all. We simply fired at their lines by guess, and occasionally the blaze of their guns showed exactly where they stood." Their Confederate opponents "kept their line like a wall of fire." The soldier hoped to train his fire at a specific target but found it difficult: "When I fired my first shot I had resolved to aim at somebody or something as long as I could see, and a dozen times I tried to bring down an officer I dimly saw on a gray horse before me."[78] In the main, the volley fire exchanged from linear formations struck soldiers in the line like raindrops in a storm, with a similarly arbitrary effect. Within the line, who was struck and who was spared often seemed random — a matter of luck, fate, or divine providence.[79] In a vivid letter to his wife, Civil War soldier Hillory Shifflet described how volley fire killed one soldier in the line while leaving the very next unharmed. His comrade George Ennis was "shot through the heart and dide instantly." But Shifflet, standing "right clos to him," emerged without a scratch though with a host of close calls: "My cap box was shot off of my belt and fore bullets holes in my blouse and my gun barrel shot off at the brech."[80] In the dispersed system, by contrast, more and more soldiers came to feel that enemy troops were targeting them specifically. Rather than being struck randomly amid a crowd of similarly vulnerable comrades, the isolated soldier often felt as though enemy gunners were trying to kill him personally. As a GI on the beach at Omaha recalled, "I felt as though every German that was on shore was aiming at me. I was the only one they were shooting at."[81] The isolation of the empty battlefield helped encourage that sensation of being singled out and targeted specifically; dispersion brought an increased focus to the individual. As one twentieth-century warrior put it, "These bullets were aimed at me. They were meant to kill me personally." To the soldier on the receiving end, there was a terrifying sense of being caught in an opponent's crosshairs: "No careless indiscriminacy about them."[82] That focus changed the way soldiers thought about the danger of the battlefield, and the way they responded to those dangers. In the words of one World War II soldier, "You take things damned personally when you're being shot at."[83]

Sometimes, of course, those soldiers were in fact targeted specifically. Perhaps the clearest example was the widespread use of snipers, carefully hidden and outfitted with specialized rifles and powerful scopes. Those marksmen sought out specific targets and frequently killed with a single, devastatingly accurate shot. Soldiers who watched a squadmate felled by a sniper's bullet received powerful reinforcement for the notion that modern warfare singled out specific victims. But despite the fact that GIs in the dispersed system felt targeted, objectively that was not often the case. Much of the fire exchanged on the battlefields of the Second World War was untargeted suppressive fire, intended to force enemies to seek cover and thus temporarily restrict their own fire, rather than to kill them outright. Regardless of the reality, however, soldiers suffering under the fire of high-powered, devastatingly accurate weapons in the Second World War felt a sense of personal victimization. Not just the weapons but the way they were employed fueled the conclusion that the individual soldier had been singled out for attention by the enemy. Soldiers caught in an artillery barrage, for example, were often the victims of shells lobbed from miles away, a quintessentially arbitrary, impersonal form of attack. On occasions when the enemy had access to a spotter, however, a distant battery inflicted on its victims a sensation altogether unfamiliar to predecessors in the crowded ranks of eighteenth- and nineteenth-century infantry formations: being bracketed by artillery shells. Working in concert, gunners and spotters used a series of shots to gauge the range. If the first attempt fell short, and a second overshot the target, the third shell almost always found its mark. Small wonder that to "the dogface dodging mortar shells," as one World War II veteran opined, the war "is concentrated on him," and "thus battle becomes a very personal thing."[84] A soldier who waded ashore during the Normandy invasion remembered that concentrated personal focus, thinking "God, let me make it to the beach, please." Vivid evidence of the danger surrounded him as he made his way through the surf: "The further I walked, the shallower it got, and fellows were dropping all over the place." Yet the soldier retained his singular concern: "You don't think of that, you really don't; you just say, 'Please let *me* make it to the shore.'"[85]

A final change in the individual's experience of infantry combat on the empty battlefield was a matter of degree rather than kind. During the War of Independence, infantrymen experienced combat in bursts: battles lasted hours, or less, and one company's time exposed to fire might last only a fraction of an hour. Soldiers of the Civil War fought longer battles — many stretching more than one day — and an individual unit's time under fire (as with Prentiss's men in the Hornet's Nest) extended far longer as well. But even for these soldiers of the mid-nineteenth century, battle was sporadic; a regiment might expect to fight only a few days a month. The near-impossibility of coordinating troops in the dark made night attacks impractical, and the unreliability of firearms in poor weather meant that few battles occurred in the rain. Armies took to winter quarters in anticipation of the spring campaigning season, allowing exhausted troops a break from the psychological grind of the battlefield. Soldiers of the Second World War enjoyed no such respite from the danger of battle. World War II troops were exposed to danger almost continuously; in the Hürtgen, for example, battle subjected soldiers to the lethal presence of enemy shells more or less constantly. Even when shells were not exploding around them, soldiers lived with the perpetual and all too reasonable worry that an attack might begin at any instant. The danger of battle threatened twenty-four hours a day, and the length of the battles stretched from hours to days to weeks. Armies' ever-increasing size, and the ability of belligerent states to supply them with a steady stream of equipment, ammunition, and supplies, meant that battle occurred more often, and it lasted longer. Individual actions were far less decisive, and successful pursuit required repeated engagements over a shorter span of time. Civil War soldiers who waited impatiently for battle in 1861 and 1862 found themselves fighting nearly every day during the summer of 1864. But even for the Civil War soldier engaged in daily skirmishes, the rhythm of combat limited his exposure to danger. By the Second

World War, the increased tempo of operations placed combat infantrymen in harm's way nearly constantly. Where a soldier of the Civil War could expect to spend a few days a month engaged in skirmishes, and might participate in a larger, pitched battle every month or two, the infantryman of the Second World War frequently experienced round-the-clock exposure to danger for days on end. Darkness and poor weather offered World War II GIs no protection from enemy air raids or from distant artillery batteries. Fighting occurred constantly along the front; the possibility of night attack loomed persistently; and regular shelling disturbed sleep and rendered every part of the day potentially hazardous. The threat of long-range bombardment meant that no part of the battlefield was safe. Even routine errands (fetching a can of rations, for example, or leaving a foxhole to relieve oneself at night) became potentially lethal activities.[86] The relentlessness of that threat wore on the nerves of even the most experienced and determined soldiers. As the next chapter explores, the constant exposure to danger proved acutely detrimental to soldiers' ability to manage their fears in combat.

Together, the new technologies and new tactics altered the patterns of experience in ground combat profoundly, effectively inverting the relationship between soldiers' awareness of danger and the options available to respond to it. Locked in their massed formations, infantry soldiers in the linear system had to muster the willpower to stand and absorb the bullets thrown against them while mechanically loading and firing their weapons. They could see the source of most of the threats directed at them—the enemy line made little attempt to disguise its presence—but could do precious little to dodge that danger. After the second day at Gettysburg, a Wisconsin veteran described the barrage in vivid terms: "The air is hot with shot and shell—The smoke is almost stifling." Within the relatively close quarters of the battlefield, he could determine the precise origin of the dangers relatively easily: "As the smoke drifts away we see the Rebels retreating through and beyond the town and our boys in hot pursuit" as both sides maintained "a continuous roar of musketry."[87] Another Civil War soldier experienced the same withering effects of hostile fire at Resaca, as his unit remained stranded under the fire of Rebel cannon for what must have seemed like an eternity. Lying on a knoll, the Union troops provided the Confederate gunners an appealing target, and they watched helplessly as the rebels loaded their guns and discharged them with a flash. The shells, the soldier remembered, screamed through the air. "Sometimes," he wrote, "they would just skip over us and again they strike in the ground in front of us and bound over us covering us with dirt." One such deflected shot hit a foot soldier next to him, breaking his shoulder.[88] For those soldiers of the linear battlefield, combat represented a terrifying trial to be endured, filled with dangers they could see but not actively avoid. One Civil War soldier described the exchange of volleys as a "horrible tempest of fire," replete with the "incessant din of musketry, the ringing in one's ears, the smell and smoke of gun powder, the defiant cheers, the intensity of intellection, the desperation even at last!" After firing his second shot he heard a shout from his lieutenant, standing on a boulder and peering over the heads of the soldiers in the line. To his dying day, the soldier vowed, he would never forget the lieutenant's expression, "so fearful in its intensity," as he screamed, "*Fire, boys, fire! They are advancing!*"[89] Combat on the linear battlefield required soldiers to act as automatons, loading, firing, and standing fast even as that tempest of fire descended around them.

Soldiers of the dispersed system experienced battle in an altogether different manner. Spread out from the massed formations of earlier battlefields and frequently hidden from both friend and foe, those soldiers enjoyed a great deal more latitude to act independently, but they often possessed little clear sense of exactly what they ought to do to minimize the danger in a given situation. For them, combat was not just a trial to be endured but also a puzzle to be solved. The scarcity of clear-cut danger cues on the twentieth-century battlefield, and the infrequent opportunities to absorb them, demanded shrewder and more active interpretation on the part of the average rifleman. Given the greater distances involved and opponents' use of cover and concealment, it was no longer easy to discern precisely where a particular danger originated. An American soldier at Wake Island found himself in the middle of such a deluge of automatic fire, unable to determine if the enemy slugs pocking the ground around him were getting closer or further away. "War is the shits!" he concluded. "Here I was, sitting right in the middle of one and I still didn't know whether someone was shooting at me."[90] Soldiers learned to piece together available information about their surroundings (where a hidden machine-gun nest likely sat, for example, or the camouflaged location of the enemy's forward-artillery observer) through an incomplete matrix of sensory data. Those who survived long enough to internalize the lessons of combat frequently described the process of deducing the sources of danger on the empty battlefield as both instinctive and intellectual. Often those techniques proved dazzling in their sophistication.

(Soldiers of the Second World War, for example, employed a sound-and-flash technique to locate the site of enemy mortars and artillery, using microphones to detect the noise of the report; technicians at an observation post used the information to triangulate the enemy position.)[91] One American GI learned to gauge the relative danger of unfamiliar territory through unusual means: experience taught him that Graves Registration units quickly buried the bodies of combatants once the battle had moved on, but the overworked burial details left the bodies of cows killed during the fighting to be interred by local farmers, who put off the job until frontline fighting had passed. Thus, "if you passed through an area of no dead cows and suddenly found yourself in an area where they were all over the place, you worried." The knowledge that communications teams strung telephone lines almost immediately in areas behind the front provided another subtle clue that a scout had outpaced his support: "When you saw no telephone wires along a road, that meant there were no more front-line units and you were in trouble. When you saw no more telephone lines and a lot of dead cows, that meant *real* trouble."[92] Unlike their predecessors, successful soldiers in the dispersed systems could not merely act as automatons, executing mechanical tasks by rote: the evolving rigors of combat required them to function autonomously, making critical decisions constantly to respond to threats they often only could guess at. Over time, technology changed the battlefield drastically, rendering soldiers more vulnerable and alone even as their ability to respond directly to the threats to their survival were blunted. Writing at the close of the nineteenth century, Ardant du Picq anticipated the aggregate effect of these changes as he imagined the difficulty of motivating soldiers to fight on the battlefields of the future: the problem was that "contest is no longer with man but with fate."[93]

Clearly, there were substantial differences in the nature of ground combat from the War of Independence to the Civil War to the Second World War. Despite important changes in the particulars of their experiences, however, an important set of commonalities connected soldiers' attitudes toward battle in each of the three eras. Perhaps most obviously, they did not want to die: whatever their other goals, individual soldiers on the battlefields of these wars wanted first and foremost to stay alive, usually desperately. World War II GI Roscoe Blunt came to that realization early in his tenure in the infantry, noting as he shipped out that his fellows were men "who had no desire to die alone in some war-ravaged country for a cause that was not — at that time — clear to any of us."[94] And though the technology of war changed, the evidence of combat's grisly ends surrounded soldiers of each era. Elisha Bostwick, a Continental who fought at the 1776 battle at White Plains, saw the ghastly carnage that resulted from that battle: a cannonball "first took off the head of Smith, a Stout heavy man and dashed it open" and struck another comrade in the bowels before it struck a sergeant and "took off the point of the hip bone." The nearness of the slaughter was especially disturbing to Bostwick, who reflected, "Oh! What a sight that was to see within a distance of six rods those men with their legs and arms and guns and packs all in a heap."[95] A Civil War soldier wrote similarly of his discovery of corpses of forty dead Confederates in a thicket: "The bodies lay among the bushes and trees just as they fell," he noted in his diary, "and were without exception shot through the head with musket balls." The following day he witnessed an even more ghastly scene: a Confederate captain "lying there with both eyes scooped out and the bridge of his nose carried away by a bullet." The ball had not killed the captain outright; delirious with misery, the victim continued to cry out, ordering a phantom company to form up and charge. The same soldier later viewed the spot where an "Ohio volunteer lay on his back, the brains oozing from a shot in the head, uttering at breathing intervals a sharp stertorous cry. He had been lying thus for thirty-six hours."[96] The aftermath of another battle left a similarly marked impression upon a GI in the European Theater. One early encounter with the carnage of battle registered as particularly stunning—an upside-down vehicle burning furiously, the bodies of six dismembered Americans scattered around it. Years later, the soldier remembered the scene in vivid detail: "The hedgerow was red with their blood, but the most sickening sight was a headless, armless, legless torso that had been thrown up in the air and was swinging to and fro, balanced on a temporary overhead telephone line." The scene ultimately overcame him, and the young foot soldier looked away, aware that he "could take only so much gore."[97]

That deep instinct for self-preservation changed little from the eighteenth century to the twentieth.[98] Because they wanted so badly to stay alive, and because the obvious danger of battle enveloped them so completely, soldiers experienced severe and often overwhelming fear as they confronted that danger. For Alfred Bell, a Civil War captain in the 39th North Carolina, fear in combat undermined his willingness to remain in the firing line. Writing to his wife, he admitted that he was always "*badly scared*" in combat, allowing in another letter that he "never wanted out of a place so badly in my life."[99] Leander Stillwell, a veteran of the fighting at Shiloh, wrote years later that he would have relinquished every bit of military glory "if I only could have been miraculously and instantaneously set down" in some peaceful place a thousand miles from the battle.[100] After the 1862 battle of Gaines' Mill, a Confederate foot soldier recalled how he "could hear on all sides the dreadful groans of the wounded and their heart piercing cries for water and assistance." Combat left an "awful scene" that the soldier hoped never to suffer again: "May I never see any more such in my life. . . . I am satisfied not to make another such charge."[101] Even soldiers who admitted some satisfaction at weathering the test of battle and witnessing the spectacle of combat were often in no hurry to return. "I feel amply repayed with haveing my life spared and with what I see out here and would'nt have missed being in the battle of Gettysburg for ten thousand dollars not a cent less," wrote one Union veteran. He was, however, "not over and about particular about being in another of the same style."[102]

A final basic continuity undergirded the combat experiences of soldiers from different wars and different eras: the unbending insistence of the military system that they go forward and fight despite the danger and their powerful fears. With precious few exceptions, human beings did not and do not want to expose themselves to the dangers of enemy shot and shell; combat demanded that they must. That is precisely what Civil War soldier David Thompson captured in the soldier's "predicament of exceptional awkwardness under fire." And yet they stayed and fought. The next chapter explores the fears soldiers encountered in battle by outlining some of the factors that enabled them to overcome it (for a span, at least) and by examining how those factors evolved over time.

2

Fear in Combat

Harry Bare waded ashore at Omaha Beach as part of the Allied invasion of France on D-Day in June 1944. As he surveyed the mangled bodies strewn across the shoreline, Bare experienced fear as a paralyzing, chilling presence that engulfed him utterly. Recalling the "men frozen in the sand, unable to move," he described his terror in vivid terms: "I could feel the cold fingers of fear grip me, and I'm sure it did with all the men."[1] Alvin Boeger, a soldier fighting in the European Theater during the Second World War, recalled that his first experience in battle left him "terrified and numb"; he remembered that he "could not move arms + legs to react" to the first attack on his company.[2] For young Continental Army lieutenant Samuel Webb, fighting a century and a half before those GIs, fear in combat manifested itself bodily as well, in a pronounced physical trembling. Two days after weathering enemy fire at the Battle of Breed's Hill in 1775, he wrote his brother "I never had such a tremor come over Me before."[3]

Some symptoms of fear in combat proved so common as to appear nearly universal for the infantrymen of these three wars. Despite the ubiquity of the symptoms, however, soldiers often encountered difficulty putting their sensations of fear into words. A World War II GI who fought through northwest Europe recalled his combat fear as a "feeling of imminent death" that appeared during the first mortar barrage he endured and remained with him, intermittently, throughout his experiences in battle. "It was," he wrote later, "something not readily describable, a feeling of impermanence." Like many combat veterans, he believed that, at its core, the terror experienced by infantrymen under fire was ineffable to those who had not experienced it firsthand: "I'd try to describe my fear, only it can't be done. You've got to feel it to know what it's like." Attempting to put the sensation into words, he characterized it as a brief feeling ("It didn't last long, this sharp pang, only between the crack and the pow" of a shell detonating nearby) but a potent one nonetheless. Despite its fleeting nature, "it was real and it was pretty rugged, as we used to say."[4] Another World War II GI found the experience of fear in battle similarly transient, but nonetheless traumatic enough that he preferred not to revisit it: his fear "evaporated quickly when the patrol ended, but I still remembered it too clearly to want to write about it."[5]

The variegated nature of fear compounded the difficulty of putting its sensations into words. As Eugene Sledge, a Marine veteran of the brutal World War II fights at Peleliu and Okinawa, put it, "Fear is many-faceted and has many subtle nuances."[6] What soldiers were afraid of changed from the eighteenth century to the twentieth, and it varied from person to person. Specific fears changed over the range of an individual soldier's experiences, as well: the things a soldier feared facing combat for the first time were usually very different than the fears triggered in combat the twenty-first time. Combatants expressed concerns of being killed, of being wounded or maimed, of being embarrassed or dishonored.[7]

Several qualities changed very little over time. The near-universal presence of fear in battle and the physical manifestations of it appear over and over in descriptions of battle. Though challenging if not impossible to describe, those cold fingers of fear grasped nearly every soldier in battle at one time or another, and the effects of that fear in combat exhibited many similar characteristics: paralyzed limbs, powerful urges to flee the battlefield, or dissolution into an ineffective mass of tremors, tics, and tears.[8] Most important was the necessity of overcoming, or at least counteracting, those effects of fear in battle: whether in the War of Independence, the Civil War, or the Second World War, soldiers who succumbed completely to their fears could not function amid the danger and confusion of the battlefield. The examination of fear is central to understanding both infantry soldiers' experiences and their motivation to fight, because those fears presented the most primal and crucial obstacle to performing in combat. Union soldier Abner Small witnessed those effects of fear in combat when he spotted a soldier paralyzed amid a fusillade of enemy fire. Attempting to make his way forward, the soldier faltered repeatedly, "bowing as if before a hurricane." The soldier made repeated attempts to go forward: "He would gather himself, gain his place in the ranks, and again drop behind." Ultimately he dropped to the ground in fright, still clutching his rifle. Even his adjutant's reproachful cry of "Coward!" could not start the soldier forward; in Small's estimation, the soldier refused to move because "he couldn't; his legs would not obey him."⁹ The tangible dangers of the battlefield—whether they arrived as solid shot, a hail of enemy Minié balls, canister rounds launched from cannon, .50-caliber slugs from an automatic machine gun, or a barrage of shrapnel from a mortar round exploding overhead—triggered powerful survival instincts in those exposed to them. As thousands of years of evolution screamed, *Run away*, some other force had to intone, *Stay and fight*. Fear created that tension between instinct and duty, and while enemy shot and shell killed or incapacitated some soldiers outright, the fear of such outcomes resulted in behaviors that compromised or destroyed the combat effectiveness of many more. In that sense, the story of motivating troops in battle is the story of the ways in which fear and its dictates were overcome.

Like other elements of the battle experience, soldiers' fear in combat exhibited changes over time. Though many of the physical symptoms of fear in combat remained similar, the way soldiers attempted to check and manage their fears evolved markedly as the introduction of new technologies and new tactics altered the character of infantry combat. The manner in which soldiers tried to manage their fears simply had to change: from the eighteenth and nineteenth centuries to the twentieth, technology altered the nature of infantry combat dramatically, transforming what soldiers had to endure and what battle demanded of them. As a result, what had worked within the linear system was no longer relevant on the empty battlefield. Soldiers and armies on the tightly packed linear battlefields of the eighteenth and nineteenth centuries could rely on a set of techniques to cope with fear that rested primarily upon the tangible presence of comrades in the firing line. The observable company of fellow soldiers provided at once an example to follow, the physical reassurance of companionship, and a powerful discouragement to cowardice, since individual actions were readily observed by others nearby. On the dispersed battlefields of the twentieth century, infantry soldiers relied upon a different set of incentives to deal with their fears long enough to fight. Those soldiers' coping mechanisms necessarily became less focused upon physical coercion, tangible support, and observing and imitating the actions of comrades, since new technologies and new tactics made it impossible to monitor fellow soldiers directly and continuously. As individual soldiers increasingly perceived battle as a more specific and more personalized event, the internalized perception that a combatant could exert some control over the combat environment (and, by extension, over his individual chances of survival) became an increasingly important technique to neutralize fear in battle. Though powerful, that perception had a dangerous flip side: as industrialized combat enlarged the scope of the battlefield, it removed many sources of danger past the ability of the individual soldier to respond to them. Frequently, the result of exposure to attacks that could not be answered was a dangerous cycle of helplessness and depression that, left unchecked, ultimately resulted in a dramatically reduced willingness to fight, if not complete incapacity. A final similarity in the patterns of experience emerged: as veterans of each conflict grudgingly noted, fear in combat occurred in situations that almost without fail frustrated soldiers' attempts to manage them.

Though the way soldiers dealt with their fears in battle changed dramatically as linear tactics gave way to dispersed tactics, the manifestations of that fear remained remarkably similar across the centuries. Soldiers in each conflict suffered some universal physiological symptoms as a result of their exposure to danger on the battlefield: quickened heartbeat, nausea (so prevalent among soldiers in combat that it became known to Civil War infantrymen as "soldier's stomach," but familiar to soldiers of every era; as a corporal at the Second World War's Battle of the Bulge remarked, "I always seemed to have a lump in my stomach" during his time in combat), as well as sweating, rapid breathing, and increased muscle tension. As generations of combat veterans learned once bullets began to whistle past, those reactions constituted an innate part of the human nervous response and could not be stifled willfully. Platoon sergeant Joe Coomer remembered the sensations of combat among members of his unit: "I guess I felt like most everyone else, pounding heart, dry mouth and sweaty palms."[10] Perhaps no physical symptom better demonstrates the uncontrollable nature of these fear symptoms than the sudden evacuation of the bladder or bowels in battle: in one survey of combat veterans, fully one in five admitted that they lost control of their bowels while under fire, and one in ten admitted to spontaneously urinating in their pants.[11] Those innate physical responses often proved most disruptive for inexperienced soldiers. As one World War II private told journalist Ernie Pyle, "There was no constipation in our outfit those first few days" in combat.[12]

A variety of phenomena, common to every battlefield regardless of time or place, triggered these innate fear symptoms. In the gunpowder age, noise proved the most common cue. The crack of gunfire activated the natural startle response, the instinctive flinch that follows a loud, sharp noise. That automatic response usually proved nearly impossible to stifle. Civil War rifleman Rice Bull recalled the difficulty of suppressing the reaction, even for experienced veterans. Describing his regiment's first exposure to an artillery barrage, he recalled how the "first shot was followed by others that hissed and shrieked over our heads and made everyone jump and duck." Such flinching was, Bull noted, "a nervous habit few ever fully overcame."[13] Like other innate reactions, those responses to the sudden noise of battle appeared regardless of place or time. Eugene Schroeder, a GI who fought in the European Theater, remembered the experience of a comrade sitting on some camouflage netting in the back of a jeep that ran over a small mine. The jeep itself sustained little damage, but the young man's reflexes were immediate and involuntary; as he told Schroeder, "Man, when that thing went off, my asshole sucked up about three yards of that camouflage net."[14] A Union artillerist whose brother had apparently given him some teasing about the way the "nois of the canon" made the soldier "week in the legs" wrote back defiantly, "If you could hear a shell there not knowing what it was it would make you weak in the legs or somewhere else."[15] Infantryman Harry Arnold was with his platoon when it received several "devastating blasts" of German machine-gun fire at close range; he later recalled how "guts recoiled and minds froze" amid the barrage of noise and lead.[16] Repeated exposure to these loud signals conditioned soldiers to react to similar sounds almost involuntarily. Bill Mauldin's cartoon combat buddies, Willie and Joe, exhibited one such ingrained response in a panel depicting a soldier casually unfastening his jacket as two desperate infantrymen dove for cover behind him. Oblivious, the offending soldier mused absentmindedly over his shoulder, "Ever notice th' funny sound these zippers make, Willie?"[17]

The immediate, bodily symptoms of fear anticipated a second, more cerebral response to the danger of combat: the anxiety and dread that attended a soldier's realization that his life was in mortal peril. That anxiety often peaked during periods of inactivity, when the individual soldier had ample opportunity to reflect on his circumstances; generations of soldiers in different eras and conflicts have identified the experience of waiting to receive enemy fire they know is imminent as among the most difficult sensations to endure. A New York volunteer characterized the difficulty of the wait before a Civil War battle as "a trying time for even veteran soldiers, almost unendurable for us new recruits."[18] Meditating on the danger of impending combat aggravated the soldier's fears. Ernie Pyle, waiting to go ashore in the invasion of Okinawa in 1944, described the palpable, miserable sensation in vivid

terms: "A heavy weight was on my heart. There's nothing whatever
romantic in knowing that an hour from now you may be dead."[19] A
GI fighting through Northwest Europe remembered how he and his
fellows had endowed the Germans "with miraculous powers of seeing
us through miles of hedgerows" as they patrolled an enemy-occupied
area. Such imaginings exacerbated the soldiers' already considerable
fears. "It's nasty," the soldier recalled, "to walk down a long road with
the feeling that death has his eye on you."[20] Yet another GI in battle
found himself picturing the effect his death would have on his family: "I thought of my mother—how she would react to my death."
The image proved so powerful the soldier could even visualize the
banner that would commemorate his demise: "I *saw* a gold star in her
window."[21] Another GI in the European Theater found that his fear
led to what he termed a "separation of awareness." One part, "the
judgmental and critical self," seemed to view the active self "with a
cynical and ironic humor," asking "who in the hell I thought I was
and what in the hell I thought I was doing with a rifle in my hands
stamping along through deep snow on a cold winter's night and head-
ing for a showdown with the Wehrmacht."[22] The power of the mind
to enhance the hard-wired fear responses led one Civil War soldier to
pronounce imagination the "greatest cause of cowardice." Men who
would "coolly face a visible danger," noted the nineteenth-century officer, "will stampede and disgrace themselves on some false report or
fancied terror."[23] Once plunged into combat, soldiers from each era
frequently noted that their fears underwent a transformation. Battle
demanded a concentration that crowded out many other concerns. As
a combat engineer of the Second World War put it, "You were so busy
trying to figure out what to do next that you didn't have time to be
really scared."[24] A soldier at Gettysburg described the "awful sight"
he witnessed during the battle in a letter home before adding that the
soldiers fighting there "hardly minded it at the time." "You may think
it strange," he wrote his relatives, "but after we get engaged in battle
we dont mind anything going on near us. We are looking off to sight
of the rebels."[25]

Both the physical and psychological effects of fear represented an
obstacle to performing in combat, since both made it more difficult
for the reluctant soldier to go forward. But soldiers in battle also discovered that not every consequence of fear had a negative influence
on their ability to fight. Since many of the physical symptoms of fear
were natural, biological responses intended to prepare the soldier to
deal with danger, troops exposed to the hazards of battle derived some

important benefits from their bodies' natural efforts to ready them for it. The adrenaline flooding their systems led to sharpened senses, temporarily increased muscular strength, and honed vision. The same neurochemical rush often generated a spike in awareness, a particularly useful byproduct for soldiers overcome with fatigue. Bruce Bradley, a soldier in the D-Day assault, recalled a catalog of projectiles that filled the air above his unit as they made their way to shore: "bombs and shells from the battleships, rockets whooshing overhead, ack-ack from German positions," accompanied by the bright streaks of tracers and geysers of seawater from exploding shells. The barrage triggered a surge of adrenaline in the soldiers as they headed toward the shore. For Bradley and his comrades, fatigued by lack of sleep, seasickness, and nervous hours spent bobbing in the English Channel, the boost in awareness was useful. As Bradley recalled, "I had been tired, and all of a sudden I was not tired. I was very alert."[26] Another beneficial effect of the body's fear response was the temporarily blunted effects of painful stimuli; the combination of adrenaline and other powerful neurochemicals suppressed nerves and raised the body's pain threshold, albeit only temporarily. That potent bath of endorphins helped soldiers function even when wounded and explains why some injured soldiers scarcely felt the pain from the blow that struck them. A Michigan soldier wounded during the Civil War described this counterintuitive phenomenon, noting that the first sensation of being shot was not pain but "simply one of shock, without discomfort, accompanied by a peculiar tingling," like a mild electric current. Numbness around the injured part, he noted, quickly ensued.[27] Abel Potter, an infantryman in the War of Independence, did not immediately feel the pain from the bayonet wound he suffered in battle; as he noted in his pension application a half-century later, not pain but the blood filling his boot alerted him to the gash.[28] Rice Bull described the sensation of his first wound as a Civil War infantryman as "a sharp sting in my face as though I had been struck with something that caused no pain." The blood flowing down his neck, rather than the sensation of pain itself, signaled to him that he had been hit.[29] Yet another use of fear in combat was heightened caution, steering soldiers away from needlessly reckless behavior. Fearlessness in battle, real or affected, occasionally led combatants to take foolish risks, which cost some their lives. Nearly every combat veteran had some story of a fellow soldier's cavalier behavior under fire and its graphic and unfortunate results. An infantryman with the 28th Alabama recalled that while retreating after an 1862 battle, one of his comrades walked along "very deliberately" exposed to enemy fire. Warned by others that he ought to get behind a tree, the soldier simply smiled and replied, "I am not afraid of them." An enemy ball killed him a moment later.[30] Another nineteenth-century soldier, frustrated at lying in the mud, stood up and exposed himself despite the warnings from comrades, deriding his cowering fellows by saying, "I don't think there is any more danger in standing here than lying in the mud, I have had enough of that." The soldier remained standing, leaning on his gun; scarcely a minute later a metallic sound, "as though one had taken a hammer and hit a tree with it," echoed down the line as the soldier dropped to the ground. A bullet had struck him in the forehead.[31] By elevating survival over comfort, instinctive reactions helped some soldiers refrain from costly errors in judgment.[32]

But the handful of beneficial effects that the fear impulse bestowed naturally in battle could not overshadow its core effects. For most soldiers, fear stood in the way of facing enemy fire and performing in battle. The concrete results of that fear—hysteria, paralysis, an overwhelming desire to flee—impaired their ability to function, sometimes disastrously. One GI in the European Theater found himself so terrified by particularly intense enemy fire that he could not even burrow a potentially life-saving foxhole: "I just lay there, huddled in a heap, too frightened to dig." Though he had been under fire before, "this fear was something new, something that made me shake all over as though with a chill and held me motionless, praying and cursing all at once."[33] Even for soldiers not entirely paralyzed by terror, panic sometimes frustrated critical military tasks. A D-Day paratrooper who landed in a field full of antipersonnel obstacles and small-arms fire found it difficult to assemble his rifle. "Under normal conditions," he remembered later, "I could detail-strip an M-1 rifle and take it apart and put it back together in the dark." But in the heat of battle, "I couldn't seem to get the three pieces of the rifle to stay together. I put the barrel and the stock together, and I put the trigger assembly in, and when I snapped it shut, it would come apart in three pieces again." The panicked soldier required four tries to render the weapon operational.[34]

The physical realities of fighting further aggravated the detrimental effects of fear. Heaping other typical stresses onto already overtaxed soldiers in battle left them fewer psychological reserves to hold their fears in check. Those stresses ranged from fatigue and dehydration to disease and injury. As one World War II GI put it simply, "You were always tired in combat." The exhaustion, he remembered, was a result of "poor food (if any), the fact that you slept with one edge of you always awake for a dangerous sound, the loads, the digging."[35] The fatigue so common to infantrymen, whether from lugging heavy packs, lengthy marches on worn-out soles, or the seemingly endless digging that increasingly became the ground soldier's lot as entrenchments and foxholes grew more and more critical in combat, left soldiers less able to control the effects of fear on their perceptions and reactions. Soldiers of the War of Independence and the Civil War suffered additional exhaustion from the demands of preindustrial combat, often marching through the night and proceeding directly into battle with little or no rest. Chronic sleeplessness, as familiar to the combat soldier as his uniform and his weapon, further intensified many of the sensations of fear on the battlefield.[36] Civil War soldier George Tillotson tried to convey that sense of exhaustion to his wife as he described its effects: "If you want to know just how I felt when I went into battle it would be hard telling exactly but in the first place I felt most d—d tired."[37] The recent experience of battle or anxiety over the next day's combat could steal sleep even from soldiers lucky enough to find themselves in tents on the eve of battle. Frederick Heuring, an Indiana volunteer who fought in the Civil War, remembered in his journal that the vivid experience of a day's fighting kept him awake that night long after the shooting ended: "I slept but little last night, as I was fighting the battle all over again all night and so I could not sleep."[38] That exhaustion was often even more acute for the GIs of the Second World War, who existed in a persistent state of sleep deprivation because their round-the-clock exposure to danger allowed little time for rest. One GI described the spiral of fatigue that stemmed from the constant need for vigilance: "We spent the night in a sort of half-doze. I'd stay awake until I couldn't any longer and then would wake Hershey, and so it went." The two continued to trade off, with each soldier "getting more and more groggy during our 'waking' shifts."[39]

Similarly, lack of food, and especially of water, further depressed individuals' ability to cope with the stress of fear in combat. Poor rations represented one of the most enduring gripes among soldiers, but the effects of an austere diet were not simply disagreeable. For soldiers expected to perform the physically rigorous tasks of soldiering, subsisting on short rations made it even more difficult to muster the necessary energy to face one's natural fears and fight.[40] And as unpleasant as hunger was in combat, dehydration constituted an even more severe threat to soldiers' abilities to resist the effects of their fears. A problem for soldiers in the eighteenth century, dehydration became rampant among the massive armies of the Civil War: epidemic dysentery in the ranks combined with fouled water to produce soldiers with a lingering, and often debilitating, thirst. Operating the muzzle-loading firearms of the eighteenth and nineteenth centuries further aggravated the dehydration that already plagued many soldiers. Infantry soldiers in the War of Independence and the Civil War loaded their weapons by using their teeth to tear open a paper cartridge containing the lead bullet and a pinch of gunpowder. The effects in the mouth of the paper cartridge and gunpowder, a powerful natural desiccant, further sapped soldiers of critical moisture in the heat of battle.[41] (Recalling his actions in an 1863 battle, a Civil War soldier remembered that as he bit the ends off his cartridges, "my mouth was filled with gunpowder; the thirst was intolerable.")[42] The sweat that bathed most infantrymen in combat, whether from the strenuous activity of fighting in hot climes or the nervous perspiration that accompanied battle even in the freezing European winter of 1944, served to wring out already-parched soldiers like sponges. Soldiers unfortunate enough to be struck by shot or shrapnel experienced an even more acute sensation of dehydration as they bled. Wounds aggravated thirst by activating the body's counterregulatory measures, as the bleeding soldier's system—attempting to compensate for the loss of fluid and blood volume—insistently demanded that he seek water.[43]

Exposure to the elements further aggravated the detrimental effects of fear. Soldiers forced to direct their energies toward staying warm had fewer psychological resources available to counteract their natural reactions to danger.[44] For American soldiers in the eighteenth and nineteenth centuries, exposure to the elements was most often due to the poor quality of their uniforms. The suffering of the ill-equipped Continental Army during the Valley Forge winter of 1777–1778, for example, was so acute that it rapidly passed into legend. And the quality of uniforms improved little in the decades leading up to the Civil War. Confederate soldiers frequently supplied their own homespun uniforms and suffered from chronic shortages of basic necessities like shoes. Their counterparts in the Union armies enjoyed a better supply and logistical system but still found themselves burdened with woefully shoddy goods: wooden-soled, inflexible shoes and lint blankets that melted away in the rain. Woolen uniforms stayed damp for hours when wet, and the fact that the fabric did not breathe made jackets and pants excruciatingly hot in the summer. Even the productive might of the twentieth-century American economy could not shield World War II GIs from debilitating exposure to the elements. American soldiers during the Battle of the Bulge, for example, suffered from poorly designed uniforms and insufficient cold-weather gear during the savage winter of 1944–1945. The variety of diseases with which soldiers had to contend further lowered their ability to manage fear in combat. In the armies of the eighteenth and nineteenth centuries, soldiers found that the close quarters of infantry encampments provided an ideal breeding ground for disease. Influenza and fevers spread rapidly among soldiers with low immunity, and a variety of illnesses threatened to become epidemic in camps. Even after the discovery of antibiotics, twentieth-century soldiers dealt with a plethora of other maladies: trench foot, frostbite, and, for soldiers in the Pacific, a host of tropical diseases. Because of the harsh and unforgiving environments that new technologies opened to infantry combat, World War II GIs suffered the debilitating effects of those conditions in far greater numbers than their predecessors. During winter operations in the cold and wet European Theater of Operations (ETO), for example, surgeons not uncommonly listed one in ten of a unit's casualties as victims of trench foot.[45] Bodies weakened by sickness proved even less able to deal with the staggering additional stresses of battle once they went into combat.

Amid such adverse conditions, then, it is hardly surprising that quelling mortal fears even temporarily proved a monumental task. The mechanisms by which soldiers in battle overcame their fears has proved as difficult to study as it is important to performance in battle. Charles Wilson (later Lord Moran) wrote his classic meditation on valor, *The Anatomy of Courage*, based on his observations and experiences during the First World War. That work likened courage to a bank account: each soldier, the author argued, possessed a finite supply. Exposure to traumatic events steadily drew down the account. Though some soldiers began with more and others with less, the supply was fixed—and once "withdrawn," could not be replaced.[46] World War II soldier John Babcock concurred with Wilson's assertion, adding his own analogy to describe the way repeated exposure to battle's dangers affected the men in his outfit: "Like steel that is repeatedly tempered, perhaps our hardening was rendering us more brittle, leaving us less resilient."[47] Indeed, the study of courage in war has engendered its own enormous literature. Gerald Linderman's *Embattled Courage* examined the mid-nineteenth-century soldier's definition of that elusive quality, and the ways that the butchery of Civil War battlefields led that understanding to unravel and finally to collapse. James McPherson linked the expressions of courage among Civil War volunteers to mid-nineteenth-century notions of honor in *For Cause and Comrades*.[48] John

Baynes's *Morale: A Study of Men and Courage* used the experiences of the Second Scottish Rifles at the 1915 battle at Neuve Chapelle to examine how notions of courage helped sustain troops amid the slaughter of the First World War.[49] Interestingly, given the importance of courage to soldiers' function in combat, it was not until the 1940s that an academic study of courage's opposite number, fear, emerged. John Dollard's 1943 study *Fear in Battle* surveyed 300 American veterans of the Spanish Civil War to analyze how they experienced fear in combat and how they had attempted to overcome those fears.[50] The findings he published formed the foundation for the formal study of fear in combat soldiers. Valor has often proved mysterious because it is so difficult to observe and quantify directly. Indeed, veterans and military theorists often describe courage in battle as the absence or mastery of fear. Unlike fear, however, whose symptoms frequently manifest openly (and which can, in some experimental situations, be monitored and quantified directly), courage maintains an elusive character.

Generations of veterans' experiences, however, demonstrated that it was possible to overcome the debilitating effects of fear sufficiently to function effectively in combat. Clearly soldiers found ways to surmount their fears: each of the hundreds of battles Americans have fought since the late eighteenth century represented an instance of groups of individuals overcoming their natural resistance to the danger of combat, at least for a time. The realization that it was possible to perform even amid the profound terror of combat amazed some soldiers. An infantryman at D-Day described his movement off the deadly beach, "surprised that legs and mind functioned together — I was absolutely terrified, and marveled that one could still function in such fear."[51] Some such mastery of fear clearly proved the rule, rather than the exception, over the two centuries. A Civil War infantryman remembered the very first volley his unit delivered while exposed to enemy fire. Peering down the line, "I judged that everyone felt about as I did; there was no levity now, the usual joking had ceased and a great quiet prevailed." The physical effects of fear were visible in the soldiers' countenances: "I could see pallor on every face as we brought the hammers to a full cock." Though the symptoms persisted in the men even as they readied their weapons ("I believe every arm trembled as we raised our guns to our shoulders to fire"), they were able to execute their duties nonetheless. "All eyes were to the front," he recalled, "not one looked back." It was "a testing time and there was not one of our company that did not pass the test."[52] As the combat records of regiments in the War of Independence, the Civil War, and the Second World War indicated, soldiers could and did quiet their natural responses to danger, at least temporarily.[53]

While substantial and important themes ran through the triggers, sensations, and effects of fear experienced by the infantrymen who fought in the War of Independence, the Civil War, and the Second World War, the ways that the soldiers of these wars dealt with their fears evolved dramatically over time. The changing character of battle, and the different demands it placed on soldiers in different eras, dictated those changes. Battle forced soldiers of the eighteenth and nineteenth centuries to suppress their fears altogether (or at least attempt to suppress them) in order to eliminate the telling physical signs of panic while in combat. The tactical demands of battle in the linear system made such suppression critically important among masses of soldiers packed into ranks. Experience demonstrated that behaviors could be acquired by proximity: social psychologists refer to the phenomenon as modeling, but the mechanism is familiar to anyone who has observed the behavior of people in groups. Human beings, particularly those in stressful situations, tend to mimic the behaviors of those around them, consciously or subconsciously, particularly in the absence of strong incentives to do otherwise. Emotions and behaviors could thus spread simply through observation. In the close quarters of eighteenth- and nineteenth-century linear battlefields where the actions of nearby soldiers were so obvious, a visible, physical display of fear proved especially hazardous. Such conspicuous behavior might spread as if by contagion and touch off a wave of panic among nearby troops who witnessed it, wrecking the integrity of the formation. Contemporary ideas about appropriate comportment in battle built upon the observation that behaviors could be transferred among those who witnessed them in their neighbors. Those notions held that bravery, in the form of stoicism, should be demonstrated conspicuously (especially by those in positions of leadership, the subject of chapter 4) lest visible displays of alarm touch off a disastrous flood of unchecked terror in the ranks. Garrett Watts, who served in the North Carolina militia during the War of Independence and fought in the battles of Camden and Guilford Court House, experienced the negative power of such modeled behavior in one engagement. As Watts admitted years afterward in his pension application, he was among the first to flee at the Battle of Camden in 1780. Unable to pinpoint the exact trigger that sparked his flight, he surmised that he ran because "everyone I saw was about to do the same." For Watts, the result was dramatic: rout in his unit was "instantaneous." With nearby soldiers fleeing in an obvious panic, "there was no effort to rally, no encouragement to fight."[54] An American paratrooper in World War II witnessed a similar phenomenon when thirty-eight German divisions assaulted a group of American infantrymen after a savage artillery bombardment. Noting that the situation was "total chaos," the trooper argued that "once fear strikes, it spreads like an epidemic, faster than wildfire. Once the first man runs, others soon follow." Without efforts to stem the tide, "it's all over; soon there are hordes of men running, all of them wild-eyed and driven by fear."[55]

Panic certainly constituted the most dangerous quality that could be acquired by proximity in battle, and the one that spread the most easily. But courage, too, could be demonstrated and copied in combat. A World War II GI who experienced fear-induced paralysis while under fire recalled watching his comrades respond to the danger: "Here and there men jumped up and ran ahead a few paces, then hit the ground again. Numbly I emulated the process."[56] The reminder that one's cowardice would be observed by comrades in the line further reinforced the importance of maintaining appearances on the linear battlefield. Charles Bardeen, writing a half-century after surviving a Civil War mortar barrage, remembered how the presence of a comrade, and the resultant possibility of shame and embarrassment, helped him control his fears during the bombardment. Bardeen made no secret of his own fear, nor of the fear he read clearly in a nearby comrade: "I was scared. Prest was scared; I knew he was scared, he knew I was scared; I knew he knew I was scared, and he knew I knew he was scared." Together, however, the men's fears acted in tandem to rechannel their urges to flee. If alone, Bardeen judged, either soldier "would have lost no time in getting to a place of safety," but instead of acknowledging their fears the two soldiers pretended to deliberate about what to do next.[57]

The value system that grew logically around linear tactics thus prized stoicism in the face of fire above nearly all else. Displays of fear, no matter how reasonable or legitimate, might spread rapidly and with devastating effect through the ranks. To perform in a system that required them to act as automatons, soldiers had to appear unruffled even (indeed, especially) when they were not. The behaviors that soldiers noticed and rewarded in combat reflected those necessities. One Civil War infantryman recalled seeing a maneuver that he described as "wonderful": a regiment, moving across a cleared field, was ordered to retreat; as they did so "they never turned their faces from the front but backed off the field in perfect order, firing all the way."[58] The special value that eighteenth- and nineteenth-century soldiers placed on that particular kind of outward display of valor, on the appearance that the soldier had mastered his fear, was so strong that even some soldiers who fled the battlefield in rout took pains to avoid the tell-tale symbols of disgrace. One Confederate soldier whose unit broke at Antietam remembered how his fear of bearing the physical mark of retreat equaled, for a moment, his fear of Union bullets. "Oh, how I ran!" he wrote in his journal. "I was afraid of being struck in the *back*, and I frequently turned half around so as to avoid if possible so disgraceful a wound."[59]

Direct physical coercion reinforced those expectations on the linear battlefield. In both the War of Independence and the Civil War, American armies made frequent use of threats and punishments to force their soldiers into battle. Here again the character of infantry combat aided the efforts of the military system to enforce specific behaviors. For men bunched together in tightly packed lines well within view of their officers and noncommissioned officers (NCOs), threats of physical punishment for cowardice in battle carried a great deal of weight. American units of the eighteenth and nineteenth centuries borrowed the concept of "file closers" (noncommissioned officers who marched alongside the troops and whose duties included prodding soldiers forward with the bayonet if necessary) from European armies and employed them with effect on those battlefields. Those coercive threats sought to supplant the very real danger of enemy fire with another, more proximate, fear: that of deadly, immediate reprisal from friendly troops. With infantrymen within easy view of their sergeants and officers, leaders could dispense such threats credibly. Thus the close-order tactics of the nineteenth-century battlefield allowed Confederate General Bryan Grimes to keep his North Carolinians in line at the Battle of Third Winchester even as nearby units broke by "threatening to blow the brains out of the first man who left ranks."[60] The power of those direct physical punishments created a quandary for soldiers who found the urgent, instinctive fear of enemy fire and the fear of their own military system's reprisals pulling them in two different directions. Garrett Watts, fleeing the Battle of Camden in 1780, threw away his gun and then, "reflecting I might be punished for being found without arms," picked up a drum to carry with him on his retreat.[61] Even though his army's strict discipline proved insufficient to keep Watts in the line, it remained in the back of his mind even as he violated its most important rule.

In addition, the close proximity of a soldier's comrades within the packed ranks of the linear battlefield also provided a palpable sense of physical reassurance. To judge by the frequency of those sentiments in veterans' recollections, the effects of that concrete encouragement was the single greatest factor motivating them in the face of enemy fire. As one Union veteran observed, "The man who can go out alone and fight against overwhelming odds is very rare." But "for every one such there are thousands who can 'touch the elbow' and go forward to what seems almost certain death."[62] William Thompson Lusk, an officer in the 79th New York, argued that "men fight in masses. To be brave they must be inspired by the feeling of fellowship. Shoulder must touch shoulder."[63] A fellow officer, Captain John William DeForest, concurred: "In our experience we believed that everything was lost if the men did not march shoulder to shoulder; and all through the battle we labored to keep a straight line with a single-mindedness that greatly supported our courage."[64] The power such physical reassurance wielded in battle was hardly unique to soldiers of the linear battlefield. Even World War II soldiers, forced to disperse from crowded ranks in order to stay alive, reflected on the power of such tangible closeness. Though most soldiers understood intellectually the tactical reasons that forced their separation, a World War II paratrooper recalled that "it is natural to want to be close to someone else when death reigns." As the troopers marched, one comrade remained close by, providing support: "Speer hung fairly close to my right rear, although at a respectful distance. I felt a comfort in his presence there and drew strength from him."[65]

Together, the combination of modeled behavior, direct coercion, and physical reassurance enabled most soldiers of the eighteenth and nineteenth centuries to manage their fears in combat, at least long enough to perform their duties amid the danger of battle. Stoicism, affecting the absence of fear, could cement infantry regiments together as they exchanged volleys—at least in most cases, and at least for a time. The tactics of dispersion largely deprived soldiers of that kind of literal cohesion on the battlefield: the same maneuvers that allowed them to operate within the storm of steel also erected daunting new obstacles to be overcome in order to defuse their fears sufficiently to go forward.

None of those new obstacles was more powerful than the altogether new and often overwhelming sense of isolation soldiers experienced on the empty battlefield. That isolation eroded morale not just by removing the reassuring touch of fellow soldiers but also by sowing confusion among the members of a unit. For better or worse, soldiers in the firing lines of the War of Independence and the Civil War could see exactly what was happening to their immediate neighbors. With constant monitoring an impossibility in the dispersed system, the confusion of battle invited soldiers' imaginations to fill in missing details, often negatively. Ardant du Picq referred to precisely that phenomenon in his description of what happened in open-order combat when a soldier watched a nearby comrade fall and disappear: "Who knows whether it was a bullet or the fear of advancing further that struck him?"[66] Losing a comrade to enemy fire might prove demoralizing, but the suspicion that the entire unit stood perched on the precipice of panic-induced rout represented a downright devastating conclusion for the soldier's willingness to remain. The increased pace of battle, which exposed soldiers to the danger of combat around the clock, day after day, also changed the way troops dealt with their fear. A soldier on the linear battlefield might stifle completely the body's fear responses with success for an hour or two, two or three times a month, but suppressing the symptoms of fear hour after hour for weeks at a time proved nearly impossible for most infantrymen. Writing in the late nineteenth century and noting that "man is capable of standing before only a certain amount of terror," Ardant du Picq reflected on the changes advancing technology had already wrought on the battlefield and the difficulties of maintaining the soldier's will to fight amid the ever-increasing dread on the battlefield: "Today there must be swallowed in five minutes what took an hour in earlier conflicts."[67]

The absence of the reassurance provided by the physical presence of the group and the impossibility of stifling instinctive responses continuously for long stretches of time prompted an important evolution in the way soldiers experienced and managed fear on the dispersed battlefield. One of the most pronounced changes was a shift in the value system of the battlefield: the mid-twentieth century witnessed a new set of attitudes toward the physical signs of fear. Stoicism while exposed to enemy projectiles no longer constituted the bedrock of combat performance; indeed, standing vulnerable and exposed to fire in the open had become, in most cases, suicidally foolish. And rather than cast fear as a shameful mark of weakness, as was common among earlier generations of soldiers who depended on the utter absence of visible signs of fear to keep the entire formation together, army pamphlets (buoyed by evidence from psychologists' survey data) encouraged soldiers to think of the physiological responses to fear as both natural and unremarkable. The U.S. Army's World War II pamphlet *Army Life*, for example, spoke directly to the concerns of inexperienced soldiers: "YOU'LL BE SCARED. Sure you'll be scared. Before you go into battle you'll be frightened by the uncertainty, at the thought of being killed."[68] The pamphlet's conversational tone reflected its attempt to portray battlefield anxiety as both unexceptional and entirely manageable; trembling and nervousness could be expected and were not necessarily associated with weakness or unmanliness.[69]

On the empty battlefield, where the tools used to quell fears within massed formations no longer operated with the same power, soldiers and the military systems that required them to function in the hazardous environment of battle needed to establish new ways to deal with the unavoidable stresses of combat. The dramatically new understanding of fear that emerged in the twentieth century provided one indication of how much had changed in the soldiers' environment. Nearly all of those changes eroded the traditional means that had kept soldiers fighting in earlier eras. Introducing space between soldiers removed the tangible physical reassurance of comrades and made it more difficult to observe and emulate the behaviors of the bravest and most effective. That space also made it more difficult for officers and NCOs to consistently monitor their soldiers' behavior and to dispense encouragement and credible coercive threats. That had been the critical task of leadership on the linear battlefield, where officers rarely carried shoulder arms since their function was to direct rather than to fight. Leaders on the empty battlefield could not police the behavior of all their troops constantly and consistently—especially when the necessity of utilizing cover and concealment forced soldiers to seek hidden dead spots where the unenthusiastic combatant could (as World War II General William Slim put it) flop into a shellhole and abandon the fight undetected. The waning power of traditional means to encourage soldiers to fight on the industrial battlefield led some theorists to conclude that war had finally become too terrible to wage. How could soldiers be motivated effectively when, in the words of Ardant du Picq, the contest was "no longer with a man but with fate?"[70]

Ardant du Picq numbered himself among those who believed that armies would find some means to help soldiers manage their fears long enough to fight even on the industrial battlefield. In his case, it was new methods of training and leadership that would overcome the new challenges of the dispersed battlefield; for S. L. A. Marshall, it was the "philosophy of discipline" that adjusted to the changing realities. As the increased destructive power of the weaponry necessitated "ever widening deployments in the forces of battle," Marshall noted, "the quality of the initiative in the individual has become the most praised of the military virtues."[71]

That sense of initiative sprang from a number of sources. Increasingly, effective soldiers on the dispersed battlefield came to manage their combat fears by connecting the decisions they made and the actions they took in battle with positive outcomes—most importantly, the heightened chance of their own survival. Experimental psychologists studying fear in the second half of the twentieth century established the importance of *perception* in determining the way individuals experience and cope with anxiety. The fear an individual perceives often has but a low correlation with the objective level of danger, or even with measurable physiological symptoms of fear.[72] Studies conducted during the Second World War anticipated some of these results: fliers in the Eighth Air Force, for example, reported higher levels of fear during their 1945 missions than they had during their missions in 1944, even though the 1944 missions were, on the whole, both more severe and more dangerous.[73]

Individual perception formed simultaneously one of the most important determinants in establishing fear and one of the most important ways to mitigate its consequences. Numerous studies have suggested that one of the most useful tools in overcoming the deleterious effects of fear is to muster a sense of controllability: the belief, whether objectively true or not, that the subject exerts some meaningful influence over surroundings or events. Hundreds of experiments over the past decades have demonstrated that general truth, and the finding is consistent with common observations about human nature. People respond extremely negatively to situations they perceive they cannot control. The perception of helplessness, the belief that a given outcome is likely no matter what action is taken, leads to increased levels of fear and, ultimately, to withdrawal and depression.[74]

Given the supreme randomness inherent in battle, it can be difficult to imagine soldiers perceiving anything resembling a clear sense of cause and effect amid a brutal and indifferent barrage of bullets and shells. But in fact people badly want to find evidence of cause and effect in life, not just on the battlefield. If anything, that desire to spot patterns and utilize them to establish some semblance of control is more common in people faced with situations that are uncontrollable to some degree. (Studies of baseball players, for example, have discovered that the perceived connection between rituals—wearing a lucky jersey, for instance—and success varies with the capriciousness of the individual player's position. Pitchers, whose accomplishments depend in considerable part on the fielding abilities of teammates that lie beyond their own control, are also most likely to make those connections.)[75] Many World War II soldiers, confronted with the vast and seemingly unpredictable experience of combat, attempted to convince themselves that understandable patterns did in fact exist. One GI of the Second World War believed that his hunches about battle revealed a kind of cause and effect. Searching for evidence in his own experiences, he accorded certain episodes special significance: "On a couple of occasions the place I had just left was visited with enemy fire, vindicating my hunch." On one level admitting that his decisions were "based more on luck than on reasoned military judgment," the soldier nonetheless reinforced his belief that there was a predictable and controllable pattern beneath his combat experiences. Successful predictions encouraged him "to continue to have and act on hunches that really had no supporting rationale," even as he admitted deliberately sorting his experiences, discounting those that undermined his reassuring framework: "I tried to forget a wrong guess that took me by whim from a safe spot to one visited with a flurry of mortar fire. I dismissed the error in choice by concluding simply that I zigged when I should have zagged."[76] Other accounts of battle indicate scores of ways in which soldiers perceived (or attempted to perceive) the battlefield as an environment over which they exerted some measure of control. It was, after all, an idea that most soldiers wanted to believe, since it carried such reassuring overtones. Like the World War II Marine who reassured his parents from training camp that those who perished in combat usually did so as a result of their own carelessness or a tactical blunder, soldiers who accorded power to the belief could remain confident in their personal survival because they could avoid committing those errors. The desire for a sense of controllability was hardly confined to the dispersed battlefields of the twentieth century. Foot soldiers in every era sought and appreciated evidence that through individual action they could exert some real influence over the outcome. Sam Watkins, a Confederate who fought in a number of battles in the Western Theater, remembered that he always shot at the rank and file, rather than at officers, reasoning that "it was they that did the shooting and the killing, and if I could kill or wound a private, why, my chances were so much the better"—a small thing, perhaps, but a way for a soldier to reinforce the belief that his own abilities, decisions, and actions could affect his fate.[77] Similarly, a soldier in the War of Independence remembered operating his weapon almost mindlessly in battle, firing "without thinking except that I might prevent the man opposite from killing me."[78] The notion that a soldier's actions could have a positive effect on his survival lay at the bottom of many infantrymen's behaviors over the two centuries. But with the advent of dispersed tactics and the concurrent loosening of traditional methods for managing fear, appeals to that sense of individual agency assumed a heightened importance. Heavier emphasis on controllability emerged on the empty battlefields of the twentieth century simply because something had to compensate for the diluted power of coercion and the absence of dependable physical reassurance. Soldiers in conflicts before the twentieth century looked for ways to reaffirm concepts they wanted to believe were true. But under the watchful eye of NCOs, and packed tightly alongside neighbors in a tactical formation in which opportunities to defect were severely limited and often more dangerous than staying and fighting, such beliefs provided the luxury of an extra comfort. On the empty battlefield, conditioned patterns played a far more central role in keeping soldiers committed to their duties by minimizing the apparent disparity between active participation in the fight and staying alive.

The importance of emphasizing a sense of controllability in managing fear helps explain why so many soldiers found inactivity so distressing and demoralizing. One of John Dollard's Spanish Civil War veterans reported that fear crested during periods of idleness: "The most intense fear is during a hot action when the soldier is not occupied."[79] The necessity of remaining still for long periods of time on the dispersed battlefield (lest activity inadvertently reveal a hidden position) further aggravated that distress. Ernie Pyle found that the "slow drag of those motionless daylight hours was nearly unendurable"; a lieutenant he interviewed reported that "lifeless waiting in a wheatfield was almost the worst part of the whole battle."[80] Some GIs, realizing the way that activity released muscle tension and eased nerves, took to shouting between enemy bursts, "Ya missed me, you stupid krauthead." There was, a soldier recognized, "a bit of bravado in it," but the benefits of shouting were significant: "Yelling breaks the tension and doesn't reveal very much about your position."[81] Shouting gave troops something to do, and vented accumulated stress; in addition, it provided reassurance that comrades remained nearby, even if unseen.

The desire for some assurance that combat was an environment in which the individual could exert some meaningful control also suggests one reason shelling proved such a particularly trying experience for most infantry soldiers. The depersonalization of the empty battlefield and the great distances separating assailants and target in the twentieth century left the individual soldier caught in a bombardment with precious few means to strike back at his antagonist. Artillery was, one soldier recalled, "like the finger of God," in both its power and its randomness.[82] One World War II paratrooper weathering an artillery bombardment reasoned that long-distance shelling was the cruelest form of enemy assault: "If only it had been men attacking us it wouldn't have been so bad," he recalled. "You can use rifles and knives to close in on and fight a man." Given the proper equipment, he argued, infantry could destroy even armored tanks. Long-range artillery was something else altogether. "It was hidden safely on the other side of the hill where we couldn't see it. We couldn't strike back. All we could do was lie there and be pounded mercilessly into the ground." Under long-range artillery bombardment, he and his comrades "truly were cannon fodder."[83] And the presence of buried antipersonnel mines on the battlefield (a weapon deposited by a departed and thus unanswerable opponent) wore on soldiers' psyches for similar reasons. Even the mere threat of mines had negative consequences. One GI reported that "the most discouraging thing of all was the file of grinning white skulls with the ominous words 'Achtung Minen' that lined the right side of the road for miles." Sometimes the signs were decoys ("At that time, we didn't know that there were usually more signs than mines"), but even the decoys were effective at terrorizing soldiers and undermining their morale. The knowledge that these markers were often a ruse did little to ease soldiers' concerns: "Even so, we never grew to like those signs."[84]

The shifting character of combat within the dispersed system, particularly its isolated and depersonalized nature, placed new emphasis upon perpetuating the sense that the battlefield was in some way controllable. In the absence of tangible, physical reminders, the sense of controllability (or at least predictability) helped keep soldiers focused on both the mission and their own survival. Efforts to establish a sense of controllability did not need to be objectively effective, or even realistic, in order to be helpful. One GI undergoing shelling during the Second World War remembered a nearby soldier, a man named Dobrich, who grabbed a large piece of paper that happened to be blowing across the battlefield and "pulled it over his prone body and head like a bedsheet and peeked out from under it like a turtle." One of his comrades pointed out that a piece of paper wouldn't stop shrapnel, and Dobrich agreed: "I know that, but it sure as hell makes me feel better."[85] Even such small, meaningless gestures reinforcing a soldier's belief in a controllable environment helped soldiers handle their fears in combat. Superstitions represented a way for soldiers to establish a sense of controllability, albeit a crude and limited one. A GI in the ETO recalled a favorite place to rest his hand on his weapon: "As we moved forward toward the unknown my left hand moved from its grip on the rifle below the front sling ferrule and advanced forward to grip the upper hand guard. There it felt comfortable." Though the soldier recognized that the gesture provided no real protection, he pointed out that "it is important that this bit of acknowledged superstition be stressed for, each of us in our own secret way, there was developed some idiosyncrasy of imagined comfort." The superstitious clutch of the rifle acted as a small psychological shield against the unease of impending action.[86]

But establishing such a sense of controllability could not motivate a soldier indefinitely. Given time, the grind of combat ultimately could wear down nearly every soldier exposed to its rigors. The constant activation of the body's emergency adrenaline system eroded the individual's ability to cope with stress, and the sheer randomness of battle, who was killed and who was spared, first weakened and eventually destroyed most soldiers' belief that combat constituted a predictable, controllable environment. World War II veteran Paul Fussell captured that progression neatly in what he termed the "slowly dawning and dreadful realization" that combat soldiers underwent in battle. According to Fussell, that realization usually occurred in three steps: "two stages of rationalization and one of accurate perception." The first stage resembled a kind of denial in which the soldier reasoned that death in battle "can't happen to me" because "I am too clever / agile / well-trained / good-looking / beloved / tightly-laced, etc." Gradually, that belief gave way to a slightly more realistic conviction: "It can happen to me, and I'd better be more careful." Soldiers in the second stage, Fussell argued, believed that they could somehow "avoid the danger of battle by watching more prudently the way I take cover / dig in / expose my position by firing my weapon / keep extra alert at all times, etc." Given enough exposure to combat, however, Fussell suggested that this belief would ultimately give way to "the perception that death and injury are matters more of luck than skill," producing an inevitable third realization: "It is going to happen to me, and only my not being there is going to prevent it."[87] Many soldiers, in fact, did enter combat with a feeling of invincibility like the one Fussell described. Phil Hannon, a soldier in the 106th Infantry division, remembered that a feeling of invulnerability burned in every soldier in the unit before their first battle: when an officer stood up and informed them that, over the course of the campaign, every other soldier would be wounded or killed, "We just turned to the GI next to us and said, 'That's tough shit for you, fella.'"[88] Feelings of invincibility frequently gave way to a profound appreciation for the danger of combat as infantrymen drew closer to the battlefield. There, at least initially, most soldiers clung to the hope that their actions and acumen might positively affect their chances for survival.

That sense of control, however limited, tended to maximize infantrymen's ability to manage fear in combat, since it produced alert soldiers motivated by the belief that their decisions and actions could help see them though the danger of combat. In battle, the soldier's belief that he could exert some influence over the outcome helped counteract what Ardant du Picq described as the demoralizing suspicion that warfare in the industrial age had become a struggle not with man but with fate. It provided the soldier with a sense, realistic or not, that there were ways one might wrest some control back from indifferent fate. Soldiers exposed to the often-overwhelming trauma of battle wanted desperately to believe in that possibility. As one GI of the Second World War put it, "Luck and fate be damned. If there was even the slightest chance to improve the odds, take it." His reflections indicated a pronounced willingness to be convinced that individual actions mattered in determining combat outcomes: "If a bullet missed by inches, maybe it was because you chose cover a little more carefully; if shrapnel chewed up a bush nearby, maybe it was because experience had pointed you to a safer place."[89]

But few experienced soldiers managed to sustain feelings of invincibility or controllability indefinitely, especially in the face of combat's unavoidable realities. Battle simply proved too arbitrary to maintain such a straightforward sense of cause and effect. And when the soldier's sense that he could exert some meaningful control over his outcome began to falter, it often sparked a disastrous cascade of mutually reinforcing perceptions that allowed the paralyzing sensation of fear to flood back in a rush. Harry Martin, a soldier at the Battle of the Bulge who was confronted with a pack of Germans threatening to overrun his position, recalled the tumble of emotions that the depressing realization stirred. Once he judged that "there was no way of stopping all of them," Martin succumbed to "a feeling of utter hopelessness; I was panic-stricken." That feeling, in turn, drained Martin's energy and will: "I felt my entire life force had left my body." Martin lost hope, feeling, as many soldiers in similar situations did, that he was already dead, "like a zombie." As panic engulfed him and his sense of controllability vanished, Martin succumbed to his fear: "I was unaware of my body, just terror." Despite his debilitating fear, Private Martin continued to take constructive action—firing, as he later remembered, "as fast as my finger could pull the trigger." But his assessment of his own effectiveness was filtered through a larger sense of his personal helplessness and served ultimately to reinforce his resignation. When the enemy kept coming on, Martin concluded that "apparently, I was not hitting a thing."[90] The reassuring power of the belief that a soldier could positively affect his outcome waned as soldiers confronted unavoidable evidence of combat's arbitrary nature. A World War II soldier faced that truth as he came upon the body of his unit's first combat victim: "Sergeant Coleman, whose professionalism convinced me he was possibly the most invulnerable soldier in the company."[91]

Flagging belief in the notion that a given action in combat could produce some effect, and the sensations of fear that suspicion encouraged, signaled the onset of fatalism. Within the ranks of the linear battlefield, as later chapters illustrate, an infantryman's sense of resignation to the outcome of battle—whatever it might be—was sometimes useful. With fewer opportunities to defect from combat and clearer and more immediate penalties for fleeing the line, a sense that one's personal outcome lay outside individual hands actually helped some soldiers of the Continental Army and the Civil War armies cope with the trials of battle. Thomas H. Evans, a Union soldier who fought at Gaines' Mill in June 1862, recalled after the war that his sense of fatalism helped him maintain his stoicism in the face of enemy fire. "There is no use trying to dodge shot," he wrote afterward, explaining that "no one hears the whistle of the ball that hits him, any more than a man sees the flash of the lightening that kills him."[92] In those armies, of course, that sense of fatalism frequently encompassed a strong religious component, a sense of predestination that saw the influence of providence in what happened or did not happen in battle. That was the explanation the chaplain of the 25th Indiana Regiment offered for his safe passage through the danger of one early combat, an explanation noticeably detached from the actions he himself took: "Although bombs and other missiles fell thick and fast around me, still through the mercy of God I am still alive."[93] David Beem, an officer in another Indiana regiment, reflected on his fears to his fiancée in a letter early in his tenure as a soldier. While the captain allowed that he had been in some danger, he reassured his sweetheart that "I never felt like being afraid in the least." The reason was his steadfast trust in God: "It seems to me the good Lord will keep me safe, so I do not fear evil. . . . As far as the danger is concerned, I have no fear, believing that a kind Providence will take care of me."[94] Their sentiments were similar to those of Continental soldier Josiah Atkins, who took comfort in the belief that "Thou, Lord, directest every ball, that none can wound unless by thy permission" and found in that faith the hope that *"tho' thousands sho'd fall at my side, & ten thousands at my right hand,* yet thou canst protect me!"[95] Yet another Union soldier wrote that he depended on his faith to help him accept the possibility of death in combat, telling relatives in 1862 that "if it should be my fate to die on the battle field I will try to submit to it cheerfully and say let God's will be done."[96] Such sentiments deemphasized the soldier's role in shaping his own fate.

Twentieth-century soldiers often described a sense of fatalism that was more secular in nature. Though many nurtured the belief that a benevolent God watched over the battlefield directing individual outcomes, many others found a degree of comfort in the simple understanding that "one has your name on it." Soldiers who had seen some combat frequently evinced a belief that death came only when one's number came up in the enormous lottery of battle. A soldier whose ticket had not yet been drawn would not be killed in any event, no matter how close death seemed to pass; conversely, even the most careful soldier could not escape harm once his time ran out. Such a belief helped soldiers on the empty battlefield manage their fears, as well. For Harold Gordon, this sense was "what made me able to carry on despite being frightened to death." His belief in a predetermined outcome helped steel him to his duty, as well: since "if you were going to get it, you would," he reasoned, "you might as well carry on with the job, as it was as safe to do that as it was to cower in a hole." Experience had taught that the danger of the battlefield was unfailingly random: "Many a time the guy in the hole had gotten it while the guy standing out in the open seemed miraculously to escape."[97] Edgar Schroeder's experiences during the 1944 invasion of France provided similar instruction. Years later the GI could recite a catalog of close calls and near misses: "One of our officers took a round through the side of his helmet, and instead of going through his head, it went up and over between the liner and the outer steel helmet, clipping his ear on the opposite side." It was not the officer's only scrape with mortality: "The same guy threw a grenade that hit a tree and bounced in his lap—a dud." These were the kinds of dramatic scenes that remained with a foot soldier, etched into his memory and shared with others later as anecdotes. Schroeder recalled other near misses. In one case, a group of GIs in a jeep whose trailer was loaded with 105-millimeter rounds sustained a direct hit on the trailer. The enemy shell ignited the propellant of the rounds in the trailer, which in turn ignited the propellant of the rounds on the hood, but—incredibly—did not detonate the cache of explosives in their payloads. "The occupants," Schroeder remembered, "suffered some scorched eyebrows and some soiled laundry." It was through these lessons that Schroeder "became a firm believer that if it ain't your time, you ain't going to get it, no matter what."[98] One Union soldier's three-part motto illustrated the tension between resignation and individual agency quite neatly: "I put my trust in God, keep my powder dry, and fear no evil."[99] The overall outcome remained in the hands of fate, but the soldier nevertheless took pains to fulfill the small functions that might maximize his chances of survival once battle was engaged.

A paradoxical tension existed between controllability and fatalism in managing fear. On one hand resignation sometimes helped soldiers deal with fear by creating a sense of fatalism that helped some cope with the anxiety of combat. A GI in the ETO who watched a squadmate make a one-man attack on a German position spent half a century puzzling over his friend's motivation for the seemingly suicidal gesture. "I think I know the answer," the soldier wrote in his memoir. His comrade "knew he was going to die, and just wanted to get it over with. Surprised that the Germans didn't do the job, he decided to see if he could make it back alive, and did."[100] Initially, resignation could help a soldier deal with the fear of death; on the other hand, total surrender to feelings of helplessness rapidly led to lethargy, depression, and, eventually, psychological breakdown.

Given enough exposure to combat, of course, soldiers in both the linear and dispersed tactical systems discovered that checking the instinctive responses to danger indefinitely proved all but impossible. Few soldiers managed to resist their fears for long without extraordinary encouragement from the military system, their training, and their fellow infantrymen. Nor did repeated exposure to fear in combat necessarily desensitize soldiers to its unpleasant effects; as one soldier admitted to a comrade, "I am not becoming any fonder of being shot at day after day. I can't get used to it. I never will get used to it." For him the sensations grew more disturbing with every episode: "The anxiety hangs on, builds, eats me up." Walking though an enemy-occupied town, the GI struggled to keep his abdominal muscles stiff, "as though that would stop a shot in the gut." Every muscle remained tense, anticipating a shot from the terrifying German 88-millimeter gun, "even my eyebrows."[101]

Though establishing a sense of personal agency was consistent with soldiers' natural inclinations, the perception itself did not arise naturally. Rather it had to be conditioned gradually over time. Much of that conditioning was deliberate, furnished carefully by a military system that taught soldiers of the mid-twentieth century to view combat in particular ways: as a place where logical cause and effect still held to some degree, and where specific actions yielded specific outcomes. Soldiers' behaviors were also conditioned by their own observations and experiences in combat; often, those experiences corroded the fragile belief that battle did indeed represent a controllable environment by underlining its unpredictability. As soldiers encountered the undeniably arbitrary destruction of combat, they moved from the second stage in Fussell's construction ("It can happen to me, and I'd better be more careful") to the third: "It is going to happen to me, and only my not being there is going to prevent it."

Nor was merely inculcating a belief that battle's outcome could be positively influenced by soldiers' decisions sufficient to guarantee their productive participation on the empty battlefield. Troops convinced of an ability to influence their own outcomes might be tempted to exercise that control to flop into Slim's shellhole, or to opt out of combat in some other unnoticeable way. To successfully channel the soldier's sense of agency toward fighting, that perception had to be aligned with military goals: soldiers had to be convinced that committed participation in battle, even in the face of the storm of steel, was not a death sentence and could actually increase their chances of survival: *do these things and you'll be alright.* For a conditioned sense of individual agency to help fulfill a unit's goals, the military system had to align the soldier's desire to live with broader military missions. Aligning a soldier's instinct for self-preservation with the army's need for troops to go forward into danger can seem paradoxical at first glance, since forward movement brought soldiers closer to enemies and the attendant danger of their weapons. But those lessons ultimately attached in many cases. A World War II veteran deftly sketched that alignment in advice he offered to fellow soldiers—advice that explicitly connected their chances for survival with forward movement by highlighting how a committed advance could actually reduce the danger: "Memo to an infantryman," he wrote, "when attacking, keep close to the enemy so he can't bring his mortars and artillery into action without endangering himself—else he will chew your asses up."[102] A soldier in the Normandy invasion exhibited a similar example of such conditioned behavior aligning the instinct for self-preservation with the necessity to press the attack. Terrified and frozen on Omaha Beach, he finally realized that only determined action would save his life: "I got my wits and realized now there's only one way to go, baby, and that's you gotta go in. And we did." Hopping from one obstacle to another, he and his comrades eventually made their way off the shell-pocked beach.[103] Tellingly, his decision to act pushed him forward, rather than leading him to halt his advance and seek shelter.[104]

Those mechanisms for influencing perceptions and behaviors in combat did not arise spontaneously. Marshall realized as much; his prescription for managing the new difficulties of the empty battlefield included training "for a higher degree of individual courage, comprehension of situation, and self-starting character in the soldier."[105] And the record of American armies in two centuries of battle indicates that the vast majority of soldiers did fight on, even in the face of the cumulative stresses that tended to drive them out of combat. Some important continuities emerged in the way these men managed their fears. Dollard's findings concluded that two of soldiers' three most important controls over fear were leadership and military training: nearly half of his veterans described leadership as a powerful way to manage fear, and more than two-fifths identified military training as similarly useful. The next two chapters discuss how soldiers' training helped establish a framework for action (Marshall's "self-starting character") and how leadership in battle attempted to reinforce those lessons — and how both changed over time in response to the evolving demands of ground combat.

3

Training

Thomas Jacobs's preparation for infantry combat during the Civil War was perhaps most notable for its mind-numbing monotony. The New York volunteer and his unit spent the majority of their training days engaged in drill. After roll call at six in the morning, the men passed the bulk of the daylight hours practicing the manual of arms (a set of detailed steps to guide infantrymen in handling, loading, and firing his musket) and rehearsing troop movements, first by companies and then in ever-larger units, most likely using exercises from William Hardee's 1855 book *Rifle and Light Infantry Tactics*. A day's training for the New Yorkers ended with two hours of dress parade from four till six in the evening.[1] The daily routine offered scarce variety; one of Jacobs's contemporaries in the 33rd Massachusetts Volunteers described his early days in the infantry as a relentless sequence of "drill, drill, battalion drill, and dress parade."[2] Drilling in camp in this manner, often alongside recruits from home, was typical for soldiers of the Continental Army and of both armies during the Civil War. The amount of drill a unit received varied depending on the value senior officers attached to it; a few eighteenth- and nineteenth-century units, particularly at the beginning of the conflicts, went into combat with no training at all.[3]

World War II GI Leon Standifer's infantry training was far more exhaustive, and as varied as Jacobs's preparations were monotonous. It was also far more comprehensive: Standifer's training was designed to build technical facility with weapons and tactics, to prepare its graduates to face the varied dangers of twentieth-century combat, and to create morale among the members of the unit. Standifer arrived at Fort Benning, Georgia (the "Fort Benning School for Boys," as he and his fellows came to call it) and was thrown in with other draftees from all over the southeastern United States. Besides marching and drilling with the M1 rifle, activities that would have been familiar to his predecessors in the armies of the War of Independence and the Civil War, Standifer and his classmates filled their days with a wide array of activities. Their days in camp were consumed with textbook tactical problems, practice exercises like Benning's "Assault on an Enemy Village," lectures from veterans on the reality of life at the front, movie screenings from the *Why We Fight* series, training in the techniques of night patrols, and occasionally esoteric barracks discussion of the morality of war in general and of killing in particular.[4]

In a general sense, the training regimens of both the linear and dispersed eras had some common purposes. Foremost was the need to take civilians unfamiliar with the military and with combat and turn them into soldiers prepared to perform amid the rigors and traumas of the battlefield. Despite obvious differences in the kinds of activity that filled soldiers' training days, combat preparation in both eras shared several goals. The first was to socialize men into the military system: to strip them (to varying degrees) of their individual autonomy, to orient them to the military's system of hierarchical authority, to prepare them to execute orders promptly and without question, and to build unity within the group. The second was to teach recruits the basic techniques of soldiering: how to load and fire their weapons, how to march, how to move from column to line or crawl beneath barbed wire as machine-gun bullets tore past overhead. These lessons also helped develop the endurance necessary to carry out the physically grueling tasks of fighting. The last was to prepare soldiers to manage their fears and to equip them with some tools that might help cope with the stresses of combat.[5]

Despite sharing some broadly similar goals, the character of infantry training changed dramatically over time. Like other facets of the infantry soldier's experience, many of those changes were in large part a response to tactical evolutions in combat.[6] The techniques taught in training changed according to the demands of soldiering in each era: combat within the linear system demanded a radically different set of skills than did dispersed combat. A company in the Continental Army, for example, needed to learn the technique of redeploying from column to line (from a marching formation to a firing formation) and to master it so thoroughly that the maneuver could be executed perfectly, even under waves of hostile fire on uneven ground. But as the massed, orderly, and suicidal German attack described in chapter 1 demonstrated so graphically, a World War II infantry company had few uses for such large-scale orchestrated maneuvers in battle. Instead, individual squads and individual soldiers would need to belly-crawl across a field menaced by machine-gun fire and move against an enemy position in independent, coordinated teams. (One measure of the increased complexity of preparing soldiers for the infantry battlefields of the mid-twentieth century is the relative length of the training manuals in use in different eras: Steuben's *Regulations* for infantry training during the War of Independence ran to 151 pages; Hardee's 1855 manual, one of the most widely used guides during the Civil War, comprised two volumes. By contrast, the American army of the Second World War produced thousands and thousands of pages of field manuals, leadership pamphlets, and battlefield guides on tactics, motivation, and morale.)

The training regimens of different eras stressed the different tasks of socialization, technical preparation, and fear management to varying degrees. In the Continental Army, for example, training exercises paid little explicit attention to the fears soldiers would experience in combat; absent an understanding of the physiological mechanisms of fear, such attention had little to offer the army in terms of maximizing combat effectiveness. Moreover, to talk about combat fears at any length might normalize (and perhaps thus encourage) them, a potentially disastrous development within close-order linear formations where fear spread quickly. Once delivered to the firing line, the perceived safety of the formation and coercive threats could keep soldiers at their task. By the Second World War, in contrast, soldiers in training underwent a set of specific processes designed to acclimatize them to the stress and confusion of combat so that they could become accustomed to their own physiological responses. Where training for the linear system focused primarily upon drill, the rote mastery of tasks performed in precise synchronization with a unit, training for the dispersed tactical system taught a more ambitious set of skills out of necessity. Soldiers still needed to operate within the military's hierarchical system, obeying orders from superiors without question, but they also needed to exercise a new degree of autonomy and initiative within that system. Officers could no longer pace the battlefield in the open, loudly urging their troops to concentrate their fire on an enemy formation; comrades would no longer stand elbow-to-elbow reinforcing the cadence of physical activity and the unity of group effort. The need for specialization in infantry combat and for flexible action on the battlefield demanded a much more thorough system of training in the twentieth century; better understanding of the mechanisms of fear gave rise to methods that tried to prepare soldiers to deal with these symptoms without sacrificing their combat effectiveness.

The progress of infantry training from the eighteenth century to the twentieth was evolutionary rather than revolutionary. Though the tactics of the dispersed battlefield bore little relationship to earlier linear battlefields, the training for infantrymen of the Second World War borrowed heavily from earlier regimens. Recruits continued to drill and march in formation, even though those skills had scant direct utility on the empty battlefields of World War II. Training in the mid-twentieth century took the rote drill of earlier eras and added to it new elements like realistic live-fire exercises and tactical problems. With those evident additions came a number of far more subtle lessons that encouraged soldiers to think about the battlefield as an environment that was controllable to some degree: to see combat as a system of choices and to detect patterns between the choices they made and their outcomes in battle. Presenting the battlefield as an environment that could be influenced by individual action, at least in part, promoted the soldier's interest in learning the activities and habits that would maximize the chances of survival. The conditioning that occurred during training helped the military system align the individual soldier's self-interest with the larger goals of the unit itself. On the dispersed battlefield, many of the effective combat motivators of the linear system had lost their power. New ones, based in part around the presumption of a controllable battlefield environment and designed to nurture the individual soldier's belief that specific actions would maximize the chances of survival, assumed their place in encouraging soldiers to fight even when their instincts urged them to flee.

For most eighteenth- and nineteenth-century recruits, the process of infantry training seemed a less discrete part of the soldiering experience than it represented for the GIs of the Second World War. For those soldiers of the Continental and Civil War armies, training marked a period of transition rather than an abrupt break with their civilian lives. Many eighteenth- and nineteenth-century Americans enlisted alongside friends and acquaintances from home, proceeding to camp surrounded by familiar faces. Soldiers in the rank-and-file frequently served under officers they recognized from town, newly commissioned leaders who in many cases had no previous military experience and who spent their evenings studying manuals like Hardee's *Infantry Tactics* in a desperate attempt to stay a step ahead of their charges.[7] That proximity and familiarity served as a bridge in the transition from civilian to military life. And many units in both the Continental and Civil War armies continued to return to encampments for additional drill even as the conflict progressed. American infantry recruits of the mid-twentieth century experienced a more dramatic break with their identities as civilians. New inductees often traveled hundreds of miles from home to be processed into the army. They received the military-issue crew cut soon after arriving at basic training (emphasizing the distinct break from home and beginning the process of stripping away individual identity and autonomy); no parallel ritual existed in the armies of the War of Independence and the Civil War. Combat, rather than the process of training for it, was the defining activity of the soldier in the minds of many eighteenth- and nineteenth-century recruits. Compared with twentieth-century conscripts, who reasonably saw in their training the beginning of a new identity as soldiers, many eighteenth- and nineteenth-century Americans placed the break with their civilian lives closer to the moment they were baptized in battle. Most eighteenth- and nineteenth-century American recruits lived in an era that familiarized them with firearms, camping out-of-doors, and the physical labor so common to soldiering far more than their twentieth-century equivalents; for those soldiers, life in an infantry camp was not nearly so different from their prior experiences as civilians.[8]

The training regimens of the eighteenth and nineteenth centuries aimed to take these men and to mold them into soldiers who could fight effectively within the linear tactical system of the day. As suggested in the snapshots of combat in the War of Independence and the Civil War provided in chapter 1, those linear tactics demanded above all concerted action among the members of an infantry unit. That unified activity, both in movement and firing, could be achieved only through reliable coordination among the members of a company. Men marched together and fired their volleys together in an effort to compensate for the limitations of the weaponry and communication on those battlefields.[9] Close-order linear tactics also ensured that the unit maintained its integrity in combat; shoulder-to-shoulder formations solidified the critical physical cohesion across rough ground or in the confusion of battle. Close contact between individual infantrymen gave leaders the best (and often only) chance to exercise a degree of command over their units, through a combination of bugle calls, drumbeats, regimental standards, and shouted instructions. Eighteenth- and nineteenth-century drill thus aimed to turn men into automatons capable of moving in lockstep and discharging their arms even amid the terror and confusion of battle. That seemed an accurate description to some officers who had studied military history: in the first year of the Civil War, a North Carolina lieutenant recalled that Napoleon had said that for "a man to be a good Soldier he must first be converted into a machine," and after some combat experience of his own, the lieutenant confessed himself in agreement.[10] On the linear battlefield, the individual soldier was effective mainly in the context of the unit; with no role for individual decision making in combat, infantry training aspired to create blocs of men who moved and fired as machines. For the unit, the most dangerous occurrence in combat was for some of its members to lose their nerve and give in to the natural instinct to flee. Besides providing soldiers with the skills to manipulate their arms and to maneuver with the unit, training in eighteenth- and nineteenth-century drill offered ways to tamp down the natural desire to run.

The rote memorization of physical tasks through seemingly endless repetition formed the mechanism that conditioned the habits and behaviors of the successful soldier in the eighteenth and nineteenth centuries. Those physical rehearsals consumed nearly all of a soldier's training time. The most obvious, and most obnoxious, quality of this drill to soldiers of the War of Independence and the Civil War was its mind-numbing sameness; the very nature of rote memorization required the near-constant performance of tasks at once exhausting and dull. Drill often seemed all-encompassing to the trainees engaged in it; as one Civil War soldier described his day in an 1862 letter to his sister, "we go on drill at 9½ come off eat dinner at 12. . . . go on drill at 2 off at 4 go on dress parade at 4½. Eat 5½."[11] Another New York volunteer of the Civil War wrote his brother that drill kept the unit "busy most of the time"; he and his comrades had "squad drill at five AM till six, dress paraid seven till eight, Co drill from nine till twelve, battalion drill from four till six PM."[12] In addition to the arduous and monotonous schedule, military drill repeatedly and unpleasantly reminded the green recruit of his own inexperience and ineptitude. As one Indiana veteran described it, soon after falling in to drill a soldier was reminded of his incompetence: "You here feel your inferiority, even the Sergeants is hollering at you to close up, Ketch step, dress to the right, and sutch the like." The close presence of comrades, meant to be reassuring in combat, further magnified some of the frustrations of drill ("the man in your reer is complaining of youre gun not being held up") and the additional irritation of this complaint might set into motion another chain of events equally as objectionable as the physical exertion itself. Perhaps the offending soldier might "make some re-mark when you will be immediately tolde by a Lieutenant to be silent in ranks or you will be put in the guard house."[13]

Though most soldiers found drill almost unbearably dull, officers of the eighteenth and nineteenth centuries embraced it as a neces-sary evil, a supremely useful if not indispensable technique for mold-ing civilians into soldiers. Preaching the virtues of drill, a New York captain attributed to it remarkable transformative powers: *"Drill and discipline,"* he argued, "will make good soldiers of any man." Drilled repetition, he maintained, was the cornerstone of combat efficacy: as long as the men were fed well, "then discipline and drill will do the rest."[14] Though citizen volunteers griped about the fatigue and monot-ony of drill and chafed at the army's insistence that they take orders most of the day, that form of training succeeded in preparing many inexperienced men for the rigors of combat, at least to a degree. Drill could not prepare soldiers for the trauma of battle, of course; regi-ments, particularly those composed of untested, green recruits broke and ran from combat not infrequently during both the War of Inde-pendence and the Civil War. But drill did help provide ways to man-age fear in battle. Moreover, no practical alternative to drill existed to prepare soldiers in the armies of the eighteenth and nineteenth centuries for combat in the linear system. Monotonous repetition of the manual of arms and various marching maneuvers conferred two benefits, one individual and one communal. For the individual, simply mastering the mechanical steps necessary for fighting demanded long periods of repetition. Firing a flintlock musket like that used by in-fantrymen in the War of Independence was a complicated process; for the Continental's equivalent in the eighteenth-century British Army, the loading sequence consisted of twenty-four separate motions, from opening the pan, emplacing the cartridge, and cocking the hammer, to filling the barrel with powder from the cartridge and ramming home the lead ball with an iron ramrod.[15] Those steps had to be mastered so reliably that a soldier could perform them in the terror of actual battle, when smoke obscured the soldier's vision and the din of combat often proved deafening, when enemy balls whistled past, and when the tremors of fear made hands shake. Repetitive training through drill represented the only dependable way to school soldiers in these manipulations. The memorization gained by endless repetition con-stituted what twentieth-century psychologists have termed an "over-learned task": an action practiced exhaustively until it became second nature, its performance embedded into the muscles so deeply it was nearly instinctive. One of the most useful benefits of an overlearned task as it applied to infantry combat is the fact that stress (which generally tended to diminish the performance of physical manipulations) actually improved performance of tasks that have been overlearned in this manner. For the soldier well practiced in the manual of arms, the chaos and attendant nerves of combat could thus improve rather than decrease his handling of a musket—an enormously valuable phenomenon in the thick of battle.[16] The troops themselves often appreciated that truth; many took pride in their mastery of the manual of arms and the intricacies of dress parade. George Rowland's New York unit was fortunate in that it had a drill sergeant who had at least five years of regular army service, unlike many units whose officers were as green as the men themselves. A letter to his father revealed the pride Rowland took in his unit's progress, boasting that the experienced drill sergeant "says he never saw a Co that took as much interest and learned as fast as we do." Given a month to drill the volunteers, the sergeant promised "he will make the best Co of us there is on the ground" despite the presence of some experienced "old regulars" in the camp.[17] The obvious satisfaction at the positive comparison with the old regulars suggests the degree to which soldiers appreciated the value of their collective skills. The group's military acumen was useful not just for the precision it displayed on the parade ground but because it heightened their effectiveness in actual combat—and that increased effectiveness might improve the group's collective chances of survival when bullets began flying.

The second benefit of drill emphasized the importance of collective effort by bonding the members of the unit together in their actions. Linear tactics made the creation of automatons through repetitive drill a practical necessity, and conditioning men to act as machines granted some important advantages in the chaos of battle. Packing the men together in orderly lines, the formations provided individual soldiers with the reassuring physical presence of their neighbors (which, as the final chapter of this book discusses, helped steel their wills to fight); the hours spent moving in concert with fellow recruits fostered a powerful, visceral sense of group unity that proved enormously helpful in combat.[18] As some soldiers in the War of Independence and the Civil War learned firsthand, physical distractions gave soldiers a useful outlet for anxiety in the heat of combat, helping to steady frayed nerves. Civil War soldier John Sherman remembered being caught in a "tempest of bullets" in one engagement. His commander's response was unusual but effective: with men falling all around, the regiment's lieutenant colonel put the men through the manual of arms. "It was a good thing for our men," he wrote later, "and kept them cool and collected."[19]

The training designed to prepare soldiers for the empty battlefield of the mid-twentieth century made use of many of the powerful rituals of training in the linear system, but added new elements to the regimen as well. The enormously increased complexity of twentieth-century infantry combat required more advanced lessons to prepare soldiers for its technical intricacies; new ideas about the nature and effects of fear in combat gave rise to exercises that sought to provide the soldier with a realistic, though controlled, sense of what the battlefield looked, sounded, and felt like. At the same time, those increasingly realistic training lessons offered the soldier a specific set of expectations about the battlefield (expectations often dashed by actual combat experience) as well as a prescribed way to think about cause-and-effect and the consequences of various actions under fire.

Certainly their training for combat registered differently in the minds of most American infantrymen of the mid-twentieth century. Unlike their predecessors in the linear system, many Americans preparing for combat in World War II underwent a discrete and standardized regime of basic training before venturing into combat. (That experience was not universal, however; thousands of American soldiers, particularly replacement troops, went directly to their units in the European Theater of Operations [ETO] and received most of their combat training in their outfits.)[20] Entry into boot camp marked a separate, and often terrifying, experience in its own right. New recruits were painfully aware that basic training marked the beginning of a life quite distinct from their civilian experiences. One new GI found that his first days in the army before training began were surprisingly relaxed, but discovered that "things began to get serious for us" when he and his comrades commenced "the training which was to mold us into a hard core fighting unit" at a Texas training camp. "The carefree days of college," he quickly realized, "were over."[21] The increasing distance between infantry training and the battlefield constituted another reason that training registered more dramatically in the minds of World War II GIs than of earlier American soldiers. Unlike Continental Army soldiers and Civil War–era infantrymen (who might go directly from a training camp to a nearby battlefield or, as in the case of some volunteers early in the War of Independence and Civil War, occasionally went directly into battle before receiving any formal drill whatsoever), twentieth-century GIs passed through a series of demarcated stages on their way to the battlefield. A newly inducted soldier generally proceeded from his hometown to one of the army's large training camps; spent weeks in basic training, and might afterward receive additional, more extensive specialized training; traveled to the battlefield across an ocean on a troopship; and arrived at a specialized depot before being assigned to a combat unit. Each stage gave the soldier the opportunity to consider his transition from civilian to soldier and reminded him at every step of the danger into which he would soon descend. That happened with one green GI, who quickly associated the acrid smell he encountered while moving through a bombed-out area of London en route to the front for the first time with a grim taste "of what lay ahead."[22]

Even with those differences, World War II infantry training did not discard all the traditions of the linear tactical system. Soldiers continued to practice the manual of arms (now with M1s rather than Brown Bess muskets or Springfield rifles, of course) and marched and paraded much like their early counterparts in eighteenth- and nineteenth-century armies. But there were important evolutions as well. Twentieth-century infantry training covered a far wider range of weaponry out of necessity; the Second World War featured an enormously varied arsenal for an effective infantry soldier to master. One typical GI described his six-month period of infantry training as "concentrated and thorough," as he learned to operate the rifle, machine gun, antipersonnel grenade, shoulder-mounted bazooka, and antitank grenade, a specialized piece of weaponry that was launched by an M1 rifle using a special blank cartridge.[23] Given that only one type of weapon was employed widely, such a varied training regimen was unnecessary in the armies of the War of Independence and the Civil War. But despite the modern weapons used in training, many of the techniques employed to teach these skills descended directly from the drill-oriented regimens of earlier centuries.

Spending endless hours practicing the manual of arms and synchronized marching, exercises clearly designed for linear tactics in use prior to the twentieth century, made little sense to many of these World War II recruits. In the pamphlet *How to Get Along in the Army*, by *"Old Sarge,"* the eponymous author offered a preemptive explanation for the necessity of long hours of practice. Expecting that soldiers would wonder why such drills were essential (especially when "common sense tells you that an enemy cannot be defeated by the manual of arms"), Old Sarge nonetheless made a case for the importance of these habits. Speaking with the voice of experience, Sarge assured the recruits that practice in the machinelike exercises of these long-established but now seemingly outmoded drills would be of great benefit to them in battle. "For one thing," Sarge advised, "your training in a little sleight of hand with the rifle is going to do wonders for your self-assurance when the time comes for you to use your weapon seriously."[24] That reference to the new soldier's "self-assurance" made explicit the importance of confidence in battle and implicitly connected overlearned tasks to a soldier's sense of effectiveness in battle. Establishing soldierly confidence through repetition and mastery harked back to the drill of eighteenth- and nineteenth-century armies; in the American army of the mid-twentieth century, training made the connection concrete, even though the specific skills themselves offered fewer benefits in battle. Like their predecessors, twentieth-century recruits would achieve mastery of those overlearned tasks by rote. The World War II army manual *Psychology for the Fighting Man* alluded to the value of overlearned tasks in its chapter "Training Makes the Soldier." "Habit formation," the pamphlet held, "is a further stage of learning. It depends on practice, experience, and repetition." Actions could not be conditioned simply by "learning in words how to perform it"; genuine physical practice was critical. Through repetition, the pamphlet asserted, "the operation of a machine gun or a rifle gets itself reduced to habit so that it becomes almost or entirely mechanical." The pamphlet employed a simple analogy to underline the process of overlearning: "Like walking. You do not have to think about putting your left foot forward after planting your right foot ahead. That is because you have walked so much. You could not get that way merely from listening to lectures on how to walk."[25]

Through sheer repetition, one soldier recalled, "We learned our rifles so well, we could literally field strip them and put them back together blindfolded."[26] And like their counterparts in the infantry units of the War of Independence and the Civil War, World War II soldiers witnessed the way that these overlearned tasks could soothe nerves amid the stress and confusion of battle. An engineer at Omaha Beach remembered surveying the chaos ashore during the first hours of the invasion; fear had overtaken many soldiers, who "simply froze and didn't do anything." Amid the paralysis, however, the rote memorization of overlearned tasks in training helped one soldier manage his terror: "I saw one GI just lying there calmly taking his M-1 apart and cleaning the sand out of it; he didn't seem to be excited at all."[27] Another GI reflected on the soothing rhythms of overlearned practice, describing the "refuge" he and his comrades found in "performing

the oft-practiced exercises of gunnery." "The repetition of training," he remarked, had paid off: even in combat, "we carried out our
tasks with practiced precision," despite the danger of battle. "To an
observer, ours looked like a simple Stateside camp maneuver" rather
than actions performed amid screaming bullets and exploding shells.[28]
A soldier fighting at Wake Island during World War II described the
utility of overlearned tasks when he suffered "the most dreaded incident any machine gunner can experience," a jam in the breech of his
weapon. Repetitive practice stripping and cleaning the gun helped the
soldier overcome the potentially paralyzing power of fear under fire:
"I thought I would freeze up," he recalled, "but much to my astonish-
ment, I was deadly calm." Rehearsing the steps to clear the breech
over and over again had prepared him to do it smoothly even under
the most trying of circumstances: "I had a screwdriver in my pocket,
and I flipped the breech open, reached down under the jammed cartridge with the point, and flipped the offending shell out slick as a
whistle. I closed the breech and pulled the activator to load another
shell, and tripped the trigger. The machine gun fired again."[29]

Old Sarge's justification of repetitive drill and the manual of arms
to individual new soldiers stressed the utility of such mastery in maintaining self-confidence. From the perspective of the group, the importance of these exercises lay in their power to build concerted effort in
battle and to enforce discipline as a habit. Samuel Stouffer's team advanced that explanation in its *American Soldier* volumes: achieving coordinated behavior in combat was of critical importance, the volumes
argued, and the obstacles to achieving it were enormous. The military
system, therefore, built in a "margin of safety" by creating an organizational structure that supported the "habits and sentiments" necessary to ensure prompt and exact execution of orders. Noting that this
system was simultaneously the source of "much that was resented by
the men in their precombat training," Stouffer's team withheld speculation as to whether close-order drill was the most efficient means
of creating this kind of automatically functioning organization. But
they expressed little doubt as to "the crucial importance of the mechanical, quasi-automatic aspects of Army operation in combat which
such training is intended to promote." The inculcated habits became,
by necessity, strong enough to operate even in the heat of battle; as
one wounded soldier put it, "You get a habit of taking orders when
you're in training so that when they tell you to do something, you
do so without hardly thinking." Stouffer's team accorded these habits
critical status, considering the obedience to orders reinforced via drill

"one of the most important determinants of behavior" in battle. "During the confusion of combat," they concluded, "in which any course is problematic and dangerous and individual judgment must operate under a tremendous handicap, the fact that one knew to whom to look for direction and orders, that such orders were forthcoming, and that the range of possible behavior was closely limited by established rule" carried enormous power to motivate.[30]

The mastery of these skills by rote remained an important component of infantry training. But the evolving nature of infantry combat, the wider range of dangers it presented and the substantial new demands it placed on combatants, meant that effective training had to do more than simply ingrain a larger number of overlearned tasks. Training for the GIs of the Second World War combined rote memorization with a host of other new exercises designed to equip them with the skills, both physical and psychological, necessary to fight effectively on the empty battlefield. By the beginning of the Second World War, studies by psychologists and sociologists had suggested that a carefully designed training program could help prepare men to withstand some of the specific rigors of infantry combat, from noise and disorientation to the paralyzing effects of gore.[31] Those ideas led to the introduction of lifelike "battle inoculation" preparation for infantry soldiers, providing recruits with realistic exposure to the sights and sounds of combat in an effort to diminish the distracting effects of the chaos once they reached actual combat. (Civil War armies staged some mock battles, but their role was more pageant than training exercise. One participant described such a sham battle in a letter to his sister: "The regiment of Infantry were blazing away at each other when a squadron of cavalry dashed around a piece of woods and charged down on them with the wildest yells." Women who had come out in carriages to watch the proceedings gave a "nice little city scream" as the artillery began its mock firing on the cavalry. Though the bogus battle did little to prepare the soldiers themselves for actual combat, the charade did have one useful consequence: the letter's author noted that "I am getting some accustomed to the smell of powder.")[32]

In the use of battle inoculation, World War II training regimens reflected the twentieth century's more nuanced understanding of combat fear. Rather than demand that soldiers stifle their fears altogether (increasingly understood to be impossible, as well as unnecessary, for most soldiers), training instead familiarized and acclimatized troops to the sensations they would experience in combat. Architects of the training system no longer treated fear as a purely counterproductive emotion: as John Dollard's veterans of the Spanish Civil War pointed out, fear in training might serve a useful purpose by motivating men "to learn those habits which will reduce danger in battle."[33]

Preparation for the demands of the empty battlefield required more than simple drilling and marching, because the dispersed system required troops to engage combat in a new way: as independent, autonomous soldiers. Rather than merely march, load, and fire in concert, soldiers of the Second World War had to be ready to negotiate a near-constant stream of judgments, decisions, and reactions amid the disorienting terror of combat. A soldier in a World War II firefight had to determine when to pop his head up to survey his surroundings, where the enemy fire peppering his unit was coming from, whether it was safer to skirt the tree line or to attack an enemy position directly. That soldier faced a host of other snap assessments that helped decide whether a given objective was fulfilled—and often whether the soldier lived or died. Infantry training was necessary to prepare soldiers to make these kinds of judgments swiftly and effectively in combat, and it had to condition them to continue forward even when their instincts urged them to stay put. GIs of the twentieth century thus engaged with their training in a manner far different from their predecessors in earlier armies. The conclusions in *Fear in Battle* gave a sense of this subtle yet profound change in the character of training. "In order to understand the value of training," Dollard argued, "the men should visualize the situation in which the training is to be used." The contrast with training for combat in the linear system was stark; company drill in the eighteenth and nineteenth centuries required only that the soldier mimic the movements of his fellows. Attempts to imagine or understand the application and utility of those movements in actual battle was far less important, and learning to make individual, independent decisions was unnecessary and even counterproductive.[34] Engagement with the lessons of training was less important in the linear system, where effective soldiers needed only master the physical steps involved in fighting: where to stand, how to reload, how to work the cocking mechanism on a musket. Whether the soldier understood the significance of those tasks was beside the point. Officers would stand close at hand to direct their actions; it was simply unnecessary for the individual soldier to understand the larger tactical plan.

One indication of this increased individual engagement with the lessons of training was the growing importance attached to target practice. Individual aim was a fairly inconsequential part of infantry training for Continental soldiers; there was little ammunition to be

spared for the luxury of target shooting, and the inaccurate smoothbore muskets of the late eighteenth century placed little premium
on marksmanship. Though the rifled muskets of the Civil War made
accurate aim a more realistic goal for most soldiers, infantry tactics
proved slower to catch up, and the familiar massed volleys of the Civil
War placed only moderately more importance on marksmanship.
Some Civil War units engaged in sporadic target practice when activity in camp grew slow (Federals more often than their Confederate
counterparts, due to the Union's advantages in producing and supply-
ing ammunition), but shooting targets in Civil War camps possessed
nothing like the importance it would assume in twentieth-century
training regimens, since linear tactics and massed volleys helped compensate for the inaccuracies of the weapons. The increasing accuracy
of the twentieth-century's shoulder arms and the disappearance of
volley fire placed a new premium on accuracy. Rifled gun barrels and
a much more plentiful supply of ammunition increased the reach of
the infantryman's shoulder arm dramatically, and by the twentieth
century infantry tactics took advantage of the precision afforded by
the technology. Accuracy gained a new currency on the battlefield,
and soldiers increasingly took pride in their marksmanship: "I became
very good with the rifle, always achieving high scores on the firing
range," wrote one GI of his training.[35] Success on the rifle range did
more than give a soldier valuable practice in a useful combat skill; it
subtly encouraged the infantryman to perceive a measurable connection between his combat aptitude and his chances of survival—and to
think more positively about his chances in battle when he performed
well in practice tasks. New emphasis on marksmanship and explicit
rewards for its mastery helped cement perceptions of the battlefield
as controllable: a better shot, the trainee could logically assume, was
more likely to survive the rigors of battle.

The increasingly fast-paced nature of combat in the dispersed system and the snap judgments it required of individual soldiers also am-
plified the importance of combat instructors who drew on battlefield
experience of their own. Book-taught drill instructors usually sufficed
to prepare soldiers for the relatively uncomplicated, though undeniably traumatic, combat maneuvers on the battlefields of the linear era.
But the evolving character of infantry combat in the mid-twentieth
century fairly demanded instructors who could bring real-world experience to bear: troops simply seemed to learn better from veterans.[36]
Of course, the army encountered substantial difficulty in arranging
veteran soldiers to act as instructors at its training camps; in many

cases, the military simply could not spare its experienced soldiers from combat areas, where their expertise was needed more desperately. Aspiring officers themselves lamented the dearth of experience among their instructors: one recruit in Officer Candidate School reflected in 1943 that "one of the shortcomings of the school, it seems to me and to many other candidates in my platoon, is the instructors' group. It does not include enough veteran officers." On occasions when a combat-experienced veteran happened to teach a course, "the men were very attentive": they immediately realized that "they were listening to a man who knew what he was talking about because he had already been there."[37] Veterans working as instructors did more than simply demonstrate the proper operation of weapons or the approved method for executing a small-unit maneuver. Equally important for the psyches of nervous trainees was veterans' firsthand insight into coping with the stress of combat itself. One trainee, worried that the strain of clearing enemy-held towns day after day would prove unbearable, listened intently as an experienced sergeant explained the reality. "You don't do it day after day," the sergeant told the assembled conscripts. "Combat is like the rest of army life: you spend most of your time waiting." That time could be spent constructively: "You clean your rifle, oil your ammunition, and try to rest," he said, drawing subtle attention to the assumption that trainees' actions would influence their fates in battle. Some periods of combat were tougher to endure than others: "When you are clearing a town or on patrol," the veteran warned them, "you are tense and scared. You begin to think nothing is going to happen; then it flares up for a few minutes and somebody is killed. Then nothing happens for a week."[38]

Other exercises rounded out the realistic battle inoculation. Soldiers passed a rigorous course of bayonet training designed to teach hand-to-hand fighting (a skill few of them would have any call to use in actual combat) and learned, as one soldier recalled, "to crawl through a simulated battlefield with machine guns firing over our heads and explosives going off around us."[39] The combination of battle inoculation and overlearned tasks helped soldiers overcome the powerful reflexive actions of fear (particularly the tendency to freeze in place) once they entered battle. One GI remembered getting off the beach at Normandy: "There was no apprehension about going through the wire and up the hill. We'd done that in training many times, so we went over the seawall and up the hill."[40] The habits established and conditioned in training thus provided a powerful way to counterbalance fear in combat. But the realism of the World War II training camp was a particular brand of realism: it occurred in a precisely con-trolled environment where the army could illustrate that the tactics and techniques it taught did indeed work effectively. Those exercises reinforced the subtle but crucial lesson: *do these things and you'll be okay.* (Of course, even the controlled realism of these exercises contained an element of danger. Working with the actual tools of war in live-fire drills raised the possibility of real injury. One soldier bound for the ETO lost a number of men from his regiment during their training; several were killed in a live-fire exercise, and yet another died in a seemingly harmless exercise when he "accidentally fell on his bayonet going over the obstacle course.")[41]

The addition of realistic "battle inoculation" was the most obvious difference between the basic training of World War II and the camp drill of the eighteenth and nineteenth centuries. But there were other significant, though more subtle, evolutions in the way the military sys-tem prepared men to face combat. Unlike the linear battlefield, with its serried ranks of soldiers moving and firing in machinelike blocs, the empty battlefield demanded soldiers who could assume initiative and perform a wider range of more complex tasks. Complicating their assignment, soldiers on the empty battlefield could no longer rely upon the "touch of the elbow" that had so aided their predecessors in bolstering their willpower to endure. Fighting within the dispersed system, soldiers of the Second World War required other means to overcome their fears in battle.

Part of the answer, as the previous chapter explained, lay in culti-vating the soldier's own sense of personal agency: building the belief that combat was in part a predictable and controllable event, and bol-stering perceptions that soldiers could learn and do things that would increase their odds of surviving battle. Battle inoculation exercises al-lowed soldiers to practice their skills in a setting that appeared real-istic but whose parameters were rigorously controlled by the military system. Successful completion of these exercises reinforced a very specific lesson (*do it our way and you'll be fine*), which would presumably aid soldiers in battle—at least until experience taught them better. For some soldiers, those appeals to individual agency and self-efficacy began even before arrival at training camp. World War II saw the ap-pearance of a genre of books designed to instruct and reassure draftees as they contemplated imminent entrance into the army. The very ex-istence of these books revealed something about the new kind of train-ing in the twentieth century; earlier soldiers received no such manuals to ease the transition from civilian to soldier. American soldiers, it seems, have always demanded an explanation for their activities to a much higher degree than soldiers in other armies (a trait Steuben first identified during the War of Independence), but the GIs of the Second World War raised these expectations to a new level. Trainees demanded to know why they did things the way they did, sometimes in exhausting detail. Thus in *How to Get Along in the Army*, Sarge explained not just the procedure of squeezing the trigger rather than pulling it (gradually increase the pressure on the trigger, he told his readers, as the rifle's sight wavered over the target) but also *why* the trigger must be squeezed rather than pulled (to avoid flinching in anticipation of the shot, and thereby throwing off the aim). As the soldier gradually increased pressure, Sarge counseled, "finally . . . the rifle will go off." The result? "Ninety-nine times out of a hundred, the bullet is in the bull's-eye": high (not to mention mathematically precise) odds for the soldier who practiced the army's lessons faithfully.[42] On occasion, such manuals could even contain a note of subversion, a nod to incoming soldiers' mistrust of military authority. Marion Hargrove's 1942 book of advice for new conscripts, *See Here, Private Hargrove*, featured an early footnote explaining the significance of the term "GI." Recognizing the two letters as the "cornerstone of your future Army vocabulary," Hargrove informed his readers that they stood for the words "Government Issue" and that "just about everything" the soldier got in the army (even the official advice) would be GI. The author took pains, however, to point out that his book "is *not* GI" even though the lessons in its pages were largely in line with official ideas.[43]

As the size of twentieth-century armies increased, the military had to fight the soldier's sense that the scale of the conflict dwarfed whatever individual contribution he could make. Old Sarge anticipated the conclusion: "You may say to yourself, 'Shucks! I'm only one guy in a couple of million or more. I'm of no importance. What happens to me or what I do isn't going to make any difference in this man's army.'" But according to Sarge, "That's where you're wrong." The answer was itself an argument for individual agency writ large: "An army is no better than its divisions; a division is no better than its regiments; a regiment is no better than its companies; and a company is no better than its men." Turning directly to the reader, Sarge continued, "That 'men' means *you*." His conclusion magnified the importance of the individual far more than had been done in the American armies of the eighteenth and nineteenth centuries: "You are really a pretty important guy," since "the example *you* set is going to have considerable weight in determining whether there will be any weak links in the piece of chain which is your company."[44] An important part of the overall message of these books stressed the reader's active intellectual participation in his own training. Old Sarge emphasized the new soldier's engagement in the training process, discouraging readers from passively absorbing information and encouraging them to think selectively about the material contained in the book's pages: "You will have to pick and choose from this book like any other," Sarge intoned. "That must be left to you."[45] Such direct talk to new conscripts sent a simple and straightforward message: the decisions *you* make as a soldier will be important. It was a subtle but pathbreaking realization without equivalent in the eighteenth- and nineteenth-century soldier's training. In traditional linear drill, simply following directions was enough to generate entirely satisfactory results; in fact, too much analysis would likely interfere with the soldier's absorption of drill, with negative consequences for both the individual and the unit as a whole.

The appeals to the soldier's sense of agency that occurred during training itself were subtle, usually nestled within more obvious lessons. World War II GIs who trained stateside received a more or less standardized regimen in basic training, but the ways in which individual camps administered the prescribed exercises and drills could vary significantly. Instructors at one camp, for example, erected rows of mock tombstones, emblazoned with memorable rhyming mottoes like "Here lies Brown the Clown, who couldn't keep his head down," to seize the attention and imagination of new recruits as they arrived at the training station.[46] The explicit message of the attention-grabbing sign was simple—*learn to keep your head down or get it blown off*—and the nature of its appeals to the new arrival's fears and sense of self-preservation were obvious and direct. The explicit message masked a much more subtle but equally important message—*there are things you can learn to do here that can save your life*—that encouraged the soldier to think about the battlefield as an environment in which logical cause-and-effect still held. In this way the phony tombstones implied a sentiment most green recruits already wanted to believe: that Brown met an early death not because he was unlucky or because his odds of survival were ridiculously slim, but because he was careless. The comforting corollary offered to the new recruit: *you* can learn to keep your head down—*you* won't be killed.[47] Sometimes that lesson appeared not subtly and implicitly in training but overtly and explicitly. Fred Arn recalled training soldiers for combat in World War II ("a damned responsible teaching assignment") and remembered making the connection between his trainees' desire to survive and the rituals practiced in camp. It became Arn's custom to make that connection as explicit as possible at the beginning of the training cycle, address-ing the recruits, "Men, no one wants to die, particularly at your age. So, here at McClellan, the more you listen, learn, and remember, the longer you'll stay alive."[48]

Similar lessons lay just beneath the surface of the twentieth cen-tury's new realistic training exercises. Basic training's barbed-wire live-fire exercise, in which trainees crawled prone underneath strands of wire while machine-gun bullets zipped past overhead, exposed the soldier to the sound of machine-gun fire and to the peculiar sensa-tion the projectiles created as they tore through the air at the same time it gave the soldier practice in the all-important belly-crawl. But the exercise taught another, subtle lesson at the same time: since the machine guns were precisely sighted above the barbed wire, no one who followed the instructions carefully would be hurt. The comfort-ing, implicit lesson: *do it like this and you'll emerge unharmed*—fright-ened, especially at first, but unharmed. A training exercise simulating a sweep through a mock-up of an enemy-held European village offered similar lessons. The exercise was a straightforward one for advanced trainees; as one soldier remembered, the drill involved live ammuni-tion and featured pop-up silhouettes of German soldiers who appeared in windows and doorways. The soldiers had a simple task: "We were supposed to spot them and fire at them before they 'killed' us."[49] The explicit goals of the exercise were clear: to provide the soldiers with experience moving through a village setting, teach them to maintain awareness, condition them to look for enemy soldiers in windows and doorways, and furnish them opportunities to practice aiming and fir-ing their weapons in a more realistic setting than the rifle range. But successful completion of the exercise taught the members of a platoon a careful lesson, as well: that if the unit was quick enough, if it spotted hostile soldiers and fired at them before the enemy shot at them, then the members of the unit would survive. The transference of such a lesson to actual combat provided the soldier with a degree of reassur-ance, and simultaneously encouraged him to channel training energies in a militarily useful direction. An assignment like the village exercise elegantly aligned the individual soldier's self-interest with the goals of the military system. Other lessons emphasized repeatedly during infantry training reinforced the logical cause-and-effect of the battle-field. Repeated instructions stressing the importance of not bunching up in a firefight (since groups of men drew enemy fire) trumpeted a sound tactical principle but also whispered, the little things *you* do in battle can improve your odds. (Of course, such live-fire exercises were not without a genuine element of danger, and in the case of the replica village, training realism proved deadly to a pair of members from the unit. Two soldiers "were goofing off in the back of the village thinking that the exercises were over for the day." Unfortunately, "one more platoon was yet to be sent through the village," and in the exchange of fire the pair were killed by stray bullets.[50] Among other things, such tragic mishaps reminded trainees of the deadly seriousness of the business in which they were engaged.)

Subtle reminders of the importance of the individual soldier's decisions in combat permeated other facets of twentieth-century infantry training as well. Lectures on the tactics and psychology of infantry warfare, unheard of in eighteenth- and nineteenth-century training, further underlined the cause-and-effect of the battlefield. In one such lecture, GIs "learned that killing the enemy was poor psychology; it made his friends mad." Wounding the enemy was a better choice: "By only wounding him, you put three men out of action: the wounded person and the two others who must carry him back to the aid station." In actual combat, of course, it was far more likely that the trainees would busy themselves laying down untargeted suppressive fire rather than carefully picking off individual targets, but a larger lesson — that combat encompassed hundreds of meaningful individual choices — encouraged the soldier to seek actions that would increase his chances of survival. The worst course, soldiers learned in the lecture, was inaction: "If you hesitate before shooting, you may become the one who dies for his country." Hesitation meant death, the soldiers learned, "not because you decided to kill or not kill, but because you didn't decide."[51] Critically, infantry training for these twentieth-century GIs conditioned them to make decisions consistent with the military's overarching goals.

On occasion, soldiers became too engaged in the training, drawing conclusions outside those prescribed by the military system. While training with the 99th Infantry Division, Jim Bowers participated in a day's training on water survival, a course he was "not too enthusiastic about." Bowers's reasoning was clear: "I figured if you went overboard in full gear you would most likely sink like a stone."[52] Musing to himself, Leon Standifer drew a lesson unlikely to meet official approval during his infantry training, reasoning that his religious qualms about killing the enemy in combat were acceptable so long as he fired his weapon, even if he didn't aim at an opponent. Standifer reasoned that "when you fire your rifle, it attracts attention and you get shot at." Therefore, "even if you fire and never hit a thing, it keeps pressure on the enemy." The conclusion he drew was not entirely inaccurate but might have made his sergeant blanch: "Marksmanship is seldom very important in combat, but noise is."[53]

Nor did instructors at training unswervingly present the lessons taught in their exercises as a panacea for every battlefield problem. That was the difference between the "school solution" and the real battlefield. Here, too, instructors could emphasize the importance of the soldier's decision making. While presenting one such school solution (the tactical manual's official answer, or "what we *should* have done," as the soldiers understood it), one instructor took care to qualify the answer. "This is the school solution," he told his charges, "but remember that it solves a school problem. Don't expect it to work on a combat problem." The officer clarified his statement, reminding students that the Infantry School illustrated a principle by devising a problem and demonstrating a specific solution. Trainees should remember "the principle of the solution rather than the specifics." While the school solution wouldn't work in combat, the principle would. That conclusion was underwhelming to the trainees (it was "so obvious that it seemed like a joke to us"), but the soldiers soon found in combat that too many OCS graduates applied the lessons directly from the book until the Germans gave them "a little advanced training."[54] The overarching message was clear: effective (from the soldier's perspective, long-lived) infantrymen needed to learn to think flexibly *within* the army's system.

It was not always possible to portray battle as a completely controllable environment in training, of course, since real-life combat was anything but. And even inexperienced recruits sometimes saw through the gaps. A demonstration of house-to-house fighting, for example, demonstrated the role of luck in a soldier's survival. As one witness described it, "You take turns running out into the street, hoping nobody shoots at you." The support of the squad made the technique effective, since "you know that the squad is standing back away from windows, watching and ready" to return fire. From the unit's point of view, the technique was reliable, but the uncontrollable element of luck remained terrifying for the individual. "Anybody who does shoot at you," the soldier concluded, "is going to be killed—you hope he misses with the first shot."[55]

Other parts of twentieth-century infantry training stressed the importance of coordination between branches of ground forces. Leon

Standifer and his comrades received a lesson in the principles of directing artillery fire from an artillery instructor; though in combat a trained artillery spotter usually directed the fire, the instructor told the men that "you should know how it is done." "The principle is simple," Standifer recalled after the war. "You have a map exactly like the one the artillery has. You place a transparent grid sheet over the part of the map covering the area to be shelled." After telephoning the coordinates to the battery, they fired a practice round to orient the spotter; the spotter informed the artillery where the practice round hit, and the duo worked in tandem to adjust the aim. In their training lesson, Standifer learned that the easiest system was to "bracket." "You put one round beyond the point, another short of it, and adjust until the guns are aimed directly on the point." Nearly as effective was the technique of "walking" the artillery: starting at one point and moving the fire forward in fifty-yard intervals. Sitting on bleachers, Standifer's comrades then received a demonstration that left him "greatly impressed" as the instructor proceeded to walk the artillery around the field, boasting that "if there were a patrol out there, I could herd them like cattle or pin them down and wipe them out." The men enjoyed the demonstration, and afterward "talked of directing artillery as if it were an orchestra."[56] (Their enthusiasm for the power of coordinated action between infantry and artillery recalled the hard-won lessons of World War I, where foot soldiers had first mastered the demanding skill of following a rolling artillery barrage closely enough to exploit its suppressive power without coming so close as to fall victim to friendly shells themselves—an early demonstration of the power of individual agency on the dispersed battlefield.) The demonstration of the artillery's precision and potency proved enormously reassuring to Standifer's training cohort, who seem not to have registered its more disturbing corollary: German artillery will be able to do the same things, raining down destruction on the trainees' own heads once they deployed in combat. Standifer's description of the exercise gave no indication that the trainees reflected on what it would be like to be on the receiving end of such an artillery barrage—in combat, a far more frequent experience than calling in artillery.

These lessons of twentieth-century training emphasized the controllable aspects of the battlefield in an attempt to keep soldiers focused on the ways they could positively influence the outcome. In many cases, the lessons proved effective, at least initially. For some infantrymen, training conferred a sense of protection in battle: though tension among the members of one soldier's green unit increased as they got close to the front, "we all felt that the training we received would pull us thru."[57] Other soldiers shared in the feeling. Aboard a transport ship bound for the ETO, one GI recently out of training camp found that his mind wandered over a variety of subjects: "Had I paid close enough attention in mine detecting school? . . . How would I respond to combat and would I measure up? . . . What was it again the major had said to look out for in German Teller mines? Where would a Butterfly mine most likely be found?"[58] That a soldier would muse over these particular concerns demonstrated the degree to which he anticipated (and wanted to anticipate) a combat experience that would be determined in large part by skill, canniness, strength, and powers of observation: qualities conceivably under the soldier's control. Those hopeful lessons of combat training rarely survived actual battle experience. But the discipline system instilled during training exerted a powerful and useful hold even on grizzled veterans. In barracks before departing for the battle at Bastogne, for example, members of the battle-tested 101st Airborne Division acted out some of the lessons of their training: "Men sat around cleaning their weapons, sharpening their trench knives, and making sure everything would function when needed." For some of these soldiers, routine escalated to fixation: "Some men took their weapons apart and cleaned them again. Some cleaned them three or four times." The utility of the repetition was twofold: besides helping prepare the weaponry for battle, it provided the men with a constructive outlet for their nervous tension.

Unlike training exercises, however, actual combat proved to be a far messier, and far less controlled, affair. The simple cause-and-effect lessons of training ("do this and you'll be okay") gradually gave way to the realization that the battlefield was arbitrary and unpredictable —and, from the perspective of a soldier conditioned to believe in a predictable environment, mercilessly unfair. A World War II GI made that connection while surveying the casualties of one combat action: "There were dead and wounded in apparently equal measure among officers, eighteen-year-old draftees, rawhide-tough ruffians, out-of-shape rookies, Regular Army sergeants, college kids, even the brash soldier-of-fortune types." Nearly half of the 200 men who had left the United States together had already become casualties, and there was no discernible pattern to explain who had emerged unscathed and who had not: "The losers seemed to be plucked by fate like Ping-Pong balls catapulted by air jet out of a glass bingo game bowl." Certainly skill did not account for the different outcomes: "As for the old distinction between the quick and the dead, a lot of the quick lay among the dead."⁵⁹ Nor did real combat obey the rules that operated during basic training. Harry Arnold soon learned that the Germans, unlike the mock opponents furnished in boot camp, were "smart, battlewise, and tricky." Live-fire drills under barbed wire had taught trainees that a soldier who stayed low to the ground would escape the streams of machine-gun bullets overhead, but the Germans heeded a different system of rules. "One of their favorite tricks," Arnold remembered, "was to fire a tracer mix a bit high, which encouraged us to move under it." Staying beneath the streaking tracers, as training had conditioned them, provided GIs with only the illusion of safety in actual combat, however. Once the GIs exposed themselves by crawling under the visible fire, "another MG would then open up with sweeps just above ground level unmarked by tracer." To soldiers unfamiliar with the German tactics, "the low fire was deadly."⁶⁰ Such experiences in battle gradually eroded soldiers' belief in the lessons infantry training attempted to inculcate by giving lie to the insistence *do it like this and you'll be fine.*

Similar lessons awaited most of the GIs who fought in the ETO and the Pacific. Real-world combat experience quickly corroded perceptions of the controllable battlefield constructed so carefully during training. Shifting his weight in line to board a troop train, a recent graduate of basic training thought back to the months of training he had received and then taught to others as a cadreman. In particular, the untried GI "wondered whether the training process would protect me in combat and guarantee my safe return." That a soldier entertained such an idea even briefly suggests that training had indeed influenced his thought dramatically; no experienced soldier imagined that *anything* could effectively guarantee a safe return from the maelstrom of the modern industrial battlefield. Years after the war his experience had crystallized the difference between the controlled battlefield of training camp and the reality he discovered in the European Theater: "There was no way I could have known while standing on that train platform in New Jersey that practically none of what I had been taught and then passed on to others applied to reality." Most of it had been "only untested and untried theory." "No one," the soldier lamented, "had informed me, in training or elsewhere, that Army manuals are best thrown out the window on the first day of combat." As he would learn in battle, "the odds of self preservation—the most basic of all human instincts—improve only if a warrior learns quickly to protect his own ass. No training manual is going to do it for him."⁶¹

Other soldiers underwent similar realizations when they arrived on the battlefield. Training proved imperfect preparation for combat; men found that no exercise could fully prepare them for the terror, danger, and confusion of the battlefield. And once in combat, many soldiers would forget some of the lessons they learned in training. Of course, for units to remain effective in battle, the army's influence over individual soldiers' behavior could not stop at the end of training camp. In battle, the presence of living reminders of the military's goals, in the form of the leadership provided by officers and noncommissioned officers, would be critical in keeping soldiers of all three wars motivated once battle began. The character of that combat leadership, the ways it evolved over time, and the reasons for those changes form the subject of the next chapter.

4

Leadership

Aldace Walker mustered into the Union Army in August 1862, and over two years of increasingly demanding fighting he learned much about what made a man an effective leader of troops in combat. After a punishing summer soldiering in Grant's 1864 Virginia campaign, Walker wrote his father to praise one specific leader, a brigade commander named Getty. The brigadier had proved himself to be "a fighting man all over," always conspicuous on the skirmish line during a fight, an officer who had mastered his fear to such a degree that he carried himself in battle as though he seemed "to think bullets are of no account." Getty's reputation for steady and selfless courage rose even higher after he survived a wound at the Wilderness to resume command of the brigade. After his return, Walker informed his father, the men "learned to almost venerate him."[1]

William Manchester, a Marine fighting in the Pacific eighty years later, wrote of another effective combat leader, a veteran sergeant, whose outward appearance in combat assumed a markedly different form but whose reputation as a courageous and effective soldier nonetheless left a distinct impression. In a memoir some believe embellished for effect, Manchester described a sergeant who was not unfailingly steady under fire, like Walker's brigadier, and who had hardly mastered his body's instinctive responses to fear in combat. During one artillery barrage, Manchester heard the noncommissioned officer (NCO) utter an obscenity and worried that the sergeant had been hit by an incoming shell. The sergeant's response quelled his concern: "He sort of smiled and said no, he had just pissed his pants." Manchester quickly discovered that it was not an isolated occurrence. "He always pissed them, he said, when things started and then he was okay." Nor did the sergeant appear particularly ashamed; as Manchester recalled, "He wasn't making any apologies about it, either." At the same moment Manchester made a discovery of his own: suddenly aware of a warm stream running down his own leg, he blurted out, "Sarge, I've pissed too," to which the sergeant grinned and replied, "Welcome to the war."[2] The exchange did nothing to diminish Manchester's respect for the sergeant as a soldier or as a leader, and it greatly reassured him about his own ability to face both the terror and danger of combat.

It is nearly impossible to imagine a successful officer in the Continental Army or in either Civil War army who routinely soiled himself in battle, much less one who enjoyed a simultaneous reputation for bravery and effectiveness under fire. That a World War II Marine could overlook such behavior as inconsequential or even natural underlines the magnitude of the changes in the way infantrymen regarded and evaluated leadership in battle over two centuries' time. Eighteenth- and nineteenth-century infantrymen defined successful leaders largely in external terms: men whose strength, bearing, and mastery of the physical signs of fear inspired the troops by example consistently received the highest praise. Conventional wisdom held that such men were born, not made. The practice of allowing the men themselves to elect their lieutenants and captains from among their own ranks aimed in part to ensure that officers whom the men respected, and whose character they knew from civilian life, won commissions and led troops into the thick of battle. Soldiers of the Second World War measured their leaders against an equally demanding yardstick, but their assessments had relatively less to do with outward appearances. A successful combat leader might not be a dynamic personality or a physically impressive individual. Competence, more than steadfastness, provided the critical metric. An effective combat leader demonstrated competence in a variety of ways: by understanding the technical tasks involved in soldiering, by mastering a range of maneuvers and equipment, by acting quickly and decisively, by keeping the men apprised of the nature and importance of their sacrifices in the overall progress of the campaign, and, above all, by demonstrating thrift with his troops' lives. Despite numerous attempts by soldiers and scholars over the decades to distill enduring "principles of leadership" that can be applied regardless of time and place, such efforts generally produce only crude results because the particulars generated by time and place are so important. In the broadest sense the job of combat leaders remained similar no matter the specifics of technology or tactics: their task involved getting human beings to expose themselves to danger, to engage in activities that violated nearly every instinct for self-preservation, and to work in concert to achieve some objective. But the manifestations of leadership varied dramatically over time as technology and tactics reshaped the fabric of infantry combat.[3]

The evolving character of ground combat explained many of these changes in the nature of combat leadership. The close quarters of the linear battlefield afforded combat leaders vastly more direct control over their troops. Officers and noncommissioned officers could keep careful watch over their men; likewise, the men could see and monitor their leaders' actions as the battle unfolded. Thus officers on the linear battlefield could employ a combination of carrot (displaying appropriate behaviors for the men to emulate and providing encouragement as they labored) and stick (threats and physical punishments for men who resisted their duty) to great effect. A reputation for physical bravery, manifested in stoicism while under fire, did more than anything to motivate the rank and file; among other things, such stoicism served as a prominent example for imitation. The dispersed tactics of twentieth-century battlefields dramatically altered the relationship between the men who fought and the men who led them into combat. Dispersal meant that troops were often hidden out of sight (not just of the enemy but of their own comrades and officers). Modeling determined stoicism in the face of hostile fire was at the same time much more dangerous (as automatic weapons filled the air with lethal metal) and much less effective, since most of the men had their own heads down much of the time and consequently could not monitor leaders' behavior consistently during a firefight. Coercive threats, too, became much more difficult to impose reliably and lost much of their motivational power as a result. Rather than rely on the power of the visible example they set to inspire the troops, leaders had to employ other techniques to motivate soldiers to carry out their duties. A broader range of responsibilities for junior officers ushered in an organized system to teach leadership skills: enlisted men no longer elected their own officers, and officers received a highly regimented training course that attempted to teach established techniques for leading men in battle. The higher turnover in personnel created by the twentieth century's increasingly destructive brand of infantry combat demanded more junior officers to lead combat units. On the fast-paced dispersed battlefield, men would have to respond to the rank, and to the system, as much as to a particular personality.

Despite important changes in the nature of combat leadership, some important commonalities persisted across the two centuries. In all three conflicts, leadership remained indispensable in directing men in battle: no matter how good their training, soldiers thrown into the maelstrom of combat behaved in similar and predictable ways. Troops under fire panicked, forgot what they had learned, and often froze up completely. Because the lessons of training possessed relatively less power to influence soldiers' behavior once battle began (the danger and confusion of combat, and the normal physiological responses to these stimuli, eroded the power of the instructions instilled by training), a successful unit required other means to keep soldiers active in combat. The U.S. Army's 1943 pamphlet *Psychology for the Fighting Man* described this natural tendency toward battle paralysis in its chapter on training. Intellectually, it suggested, the soldier in combat grasped the importance of digging a trench to provide a measure of safety from hostile fire. Indeed, the soldier "may have learned from demonstrations exactly how to go at digging it." But the rush and confusion of combat could overwhelm the lessons of training, so that "when the enemy planes come overhead, he may, in his excitement, forget what he has learned." The result for the soldier, the pamphlet concluded, could be disastrous: "In such emergencies, he is more likely to act from habit than from reasoning and sense."[4] Like World War II soldier Lester Atwell, soldiers under fire found that their fear overrode their training. Though Atwell and his comrades had been instructed repeatedly to maintain spacing between soldiers (constant shouts of "No bunching up!" and "Keep your distance!"), the noise and danger of enemy rounds activated a powerful herd instinct that overrode their instruction, and the men began to gravitate toward one another almost immediately: five minutes after being reminded, Atwell recalled, he and his comrades "began to drift together again."[5]

Because the habits conditioned through training were by themselves insufficient to keep soldiers moving once the bullets began flying, armies sought other ways to keep their soldiers' attentions focused upon the mission. In most cases, small-unit leaders, the officers and noncommissioned officers with whom the soldier had regular, face-to-face contact in squads, platoons, and companies, bore that burden. Combat leadership derived much of its power from the simple fact that people in stressful situations often want to be told what to do. Private Frank Raila, an infantryman at the Battle of the Bulge, observed that in the heat of battle, "there are moments when somebody freezes up, unable to get going when there's a mortar barrage or artillery rounds falling." Direct instruction from a leader could break this paralysis; in Raila's experience, even frozen soldiers "respond to an order from a noncom, officer, or even a private." Such an instance, Raila recognized, was "a case where leadership makes a difference."[6]

The energy that armies historically have devoted to eliminating opposing leaders in battle testifies to their enduring importance in combat. That persistent interest in targeting enemy leaders extended across the centuries, fueled by the expectation that a group of soldiers without a leader could not function in combat, as ineffective as a body without a head. As one GI in the European Theater of Operations (ETO) remarked of his German opponents, "stop the officer and you stop his unit."[7] During the War of Independence, Continental Army sharpshooters frequently targeted the shiny gorget, a decorative golden plate worn at the throats of British officers, in an oncoming body of troops, in the hopes that eliminating an officer would dissolve the integrity of the entire enemy unit. Though the ornamental gorget ultimately disappeared from the British officer's gear, the distinctive markings that designated officers in other conflicts remained an attractive target for opposing troops. During the Second World War, morbid GIs took to calling the gleaming silver bars worn on the helmets of American lieutenants and captains "aiming stakes" after their tendency to attract enemy fire; many officers elected to replace the reflective silver emblems on their helmets with identical symbols in matte black paint to help deflect the unwanted attention. In North Africa in 1942, journalist Ernie Pyle spotted a colonel wearing a brown canvas hat without any insignia at all, and noted that "officers at the front tried to look as little like officers as possible, for the enemy liked to pick them off first."[8] Awareness of the dangers inherent in leading troops trickled down to soldiers at every level. One coxswain at D-Day who picked up a one-star general in his second boatload of troops had time to reflect on his new cargo during the treacherous trip to the beach. "I remember as a kid reading cowboy and Indian stories and how the enemy liked to kill the leaders, the man in control," he recalled later. He had developed a keen appreciation for the enemy's desire to target leaders, and understood the danger his close proximity to the general invited from the German gunners on the shore. "I kept thinking, here I've got the man—with the helmet on, with the one silver star, which is extremely visible—and I kept thinking, why don't I have the guts to ask him to turn around so they won't blow this boat out of the water!"[9]

At the same time, soldiers in different eras understood that effective leadership alone could not guarantee effectiveness in combat; it formed merely one piece of a large and complex puzzle. One of the army's own World War II pamphlets acknowledged this distinction between leadership and combat effectiveness: "Leadership and morale are not synonymous," it argued, "yet they are as inseparable as the counterparts of an electrical circuit." The pamphlet sustained the metaphor to explain the relationship: "Morale is like the current—the powerful electromagnetic force—and leadership is like the conductor that guides and transmits that force to the motor." Put simply, "the state or quality of morale produced is directly related to the quality of the conductor or leader."[10]

The way this imagined conductor channeled the efforts of individual soldiers varied widely over time. As his unit delivered volleys into the British line at Cowpens, Virginia private Jeremiah Preston noted that both his fellows and the opposing line paid close attention to the behavior of their leaders. "The contest became obstinate," he recalled half a century later in his pension application, "and each party, animated by the example of its leader, nobly contended for victory." The awareness of the manner in which both lines looked toward and responded to the examples set by their leaders (particularly the way a leader's comportment had the power, in the troops' estimation, to "animate" the line) is telling, and highlights an important underlying characteristic of combat leadership within the linear system. The Virginians' adherence to their leader's example was apparently successful, as the "line maintained itself so firmly."[11]

The power of physically embodying and prominently demonstrating specific behaviors for soldiers to emulate emerged in chapter 2. Its centrality to combat leadership in the eighteenth century stemmed from the tactical realities of the battlefields of that era. The brutally straightforward nature of the linear tactical system gave rise to a correspondingly straightforward set of valued traits for combat leaders. The most significant threat to a unit's integrity on the eighteenth-century battlefield was breaking under fire, and most often panic triggered that kind of dissolution. In battle, terror spread like a contagion, especially in the close quarters of the line: with comrades shoulder-to-shoulder, readily observable by their neighbors, the actions of a few could easily trigger the rout of an entire company. Like Garrett Watts, whose flight from the 1780 Battle of Camden was mentioned in chapter 2, many frightened soldiers ran as soon as it seemed that everyone else in view was about to do the same. The effect of panic in such close quarters was, as Watts relayed, "instantaneous." The easy transmission of panic among soldiers assembled in a firing line created the eighteenth-century officer's urgent need to demonstrate appropriate behaviors, particularly stoicism, by suppressing fear and refusing to flinch under fire, providing a tangible example for his troops to imitate. The tactical realities of the linear battlefield encouraged an understanding of

combat leadership based largely upon outward appearances and personality. Soldiers expected leaders to possess a kind of flinty stoicism,
a specific brand of masculinity that could serve as an example up and
down the friendly line. Not surprisingly, then, many of the traits that
Continental Army soldiers valued most in their leaders were physical:
explicitly visible and easily identifiable signs of manhood and character. Those characteristics took a variety of forms. Private Christopher
Brandon, for example, greatly admired the physical prowess of his
lieutenant, Joseph Hughes, who was "not only a man of great personal
strength, but of remarkable fleetness on foot." Other, more subtle
traits—an erect bearing, for example—could also support an officer's
claim to leadership in combat. One Delaware private recalled in particular how "the powerful & trumpet like voice" of his commander
possessed special currency in battle, explicitly connecting those visible
attributes with the ability to counteract fear and invigorate the troops
to carry on the exhausting labor of loading and firing. The private
confessed that the power in that voice "drove fear from every bosom,
and gave new energies to every arm."[12] Resistance to pain, whether real
or affected, constituted yet another sign of ability in combat leaders,
and another way for those in command to model behaviors for their
soldiers to mimic. The high pain threshold of a junior officer inspired a
group of Continental soldiers at the battle of Cowpens. When his men
broke and fled before a British assault, a lieutenant rallied them with
drawn sword despite the fact that he had sustained a saber cut across
his right hand. With his "loud voice" he continued to urge the men
on in spite of the wound.[13] Historian Charles Royster relates an even
more graphic story of a leader affecting indifference toward injury and
coolness in the face of danger to motivate his troops: at the October 1781 battle at Yorktown, American captain Stephen Olney stepped
onto the parapet "and called out in a tone as if there was no danger,
Captain Olney's company, form here!" Stabbed with British bayonets,
Olney held part of his intestine with his hands as he gave orders but
nonetheless survived the battle.[14] Simple physical endurance also won
the respect of enlisted men. The night before the Battle of Cowpens,
Thomas Young remembered the prodigious stamina of General Daniel
Morgan: long after Young himself lay down to sleep, Morgan "was going about among the soldiers and encouraging them," promising them
that he would "crack his whip" over the British cavalry commander
in the coming battle. Young himself registered amazement at Morgan's energy, recalling, "I don't believe he slept a wink that night!"[15]
   Modeling appropriate soldierly behavior as the battle raged was

hardly the eighteenth-century leader's only responsibility in combat. The combined excitement and terror of combat made it necessary to steady troops with direct commands. During the battle itself, Thomas Young watched as Morgan galloped "along the lines, cheering the men, and telling them not to fire until we could see the whites of their eyes." Perhaps because the wind carried their voices, Young could also hear that "every officer was crying don't fire!" In an observation that anticipated Lester Atwell's experiences in the Second World War more than a century and a half later, Young noted that the officers' constant reminders were necessary, since amid the nervous excitement of battle "it was a hard matter for us to keep from" firing early, before the enemy came into range.[16]

Amid the din and chaos of battle, however, displaying stoicism and restraint was not enough to keep men at their posts. Successful leadership within the linear ranks required not only a conspicuous example but also a variety of threats, inducements, and inveiglements to keep troops at their tasks. The behavior of the aforementioned Lieutenant Hughes at Cowpens offered a succinct distillation of those strategies. When men broke and fled in the face of the British cavalry charge, Hughes employed a bevy of techniques while attempting to reform his line. Lieutenant Hughes impugned his men's sense of honor (referring to them as "d—d cowards" when they fled); issued explicit, coercive threats (an order to "halt and fight" punctuated with strikes of his unsheathed sword); and ultimately resorted to an appeal to the men's logic and sense of self-preservation, reminding them that "there is more danger in running than in fighting, and if you don't stop and fight, you will all be killed!" Hughes's commands constituted a veritable catalog of eighteenth-century combat leadership in just a handful of well-chosen sentences, and his terse instructions left an understandable impression on the men who witnessed his actions. Once he had rallied a kernel of men, "others joined them for self-protection"—an acknowledgment of the power of modeled behavior as well as the perceived reassurance of the tightly packed ranks.[17]

The broad similarities between the linear battlefields of the War of Independence and the Civil War created a number of continuities in the patterns of leadership visible there. Soldiers still fought under the watchful eye of a lieutenant or captain, who usually spent the battle pacing the line and urging the men to keep up their fire. And the close proximity of soldiers and officers on the firing line meant that Civil War officers could still use coercive threats with effect: as one Union officer at First Bull Run noted, the men at first "seemed inclined to back out," but the officers stationed themselves behind the line and "threatened to shoot the first man that turned."[18]

Civil War regiments' practice of electing their junior officers by popular ballot spoke both to the importance of preexisting local ties between the men and their officers and the powerful assumption that soldiers in the ranks were effective judges of leadership potential. The strength and importance of the local bonds between men and their leaders appeared regularly in the informal relationships many men enjoyed with their junior officers. Rice Bull recalled a moment of relaxed levity that occurred as his unit endured a storm lying in a pool of water; one of the soldiers, "who was quite a wag," offered to "swim over and tackle the Johnnies," if only the captain would give the order, presumably to the delight of the rest of the company. To Bull, such easy familiarity between officers and enlisted men was "easy to understand," given that "our Regiment was made up entirely of men from Washington County and each Company of men from the same or adjacent townships; their officers were older men, the friends of their fathers and mothers."[19] Besides the benefit of easy relationships between officers and enlisted men, such a system revealed a belief that men fought best under and beside men with whom they enjoyed a personal connection, an idea the final chapter of this book explores in more detail. The second assumption, that the men themselves were the best judges of leadership ability, appeared over and over again in the observations of soldiers from the mid-nineteenth century. Midway through the war, George Rowland described the emerging picture of leadership in his company. Rowland clearly felt that the men wanted one of his friends, Matthew, to be promoted to captain within a few months, and expected him to be commissioned shortly: "I would not be surprised. . . . We hope he will be, he is a good officer." Rowland reserved harsher words for another man in the company, warning that "Stevens is a poor plug and he ant capable of being Capt."[20] The statements revealed an implicit belief that the enlisted men themselves could best discriminate between able officers and pretenders. That notion has certainly been the logic of any number of griping soldiers throughout history, but the conviction proved especially important in the military systems of the eighteenth and nineteenth centuries. Enlisted men's opinions still held importance in the twentieth century, and certainly World War II soldiers seized every opportunity to voice their opinions of their officers, but increasingly the official designation of rank carried a special imprimatur, due in part to the increasingly complex and elaborate training that officers received.

For Civil War soldiers of the mid-nineteenth century, however, the absence of sophisticated and formal officer selection and training led to a standard of evaluation that was in many ways similar to that of the eighteenth century. Outward appearance still counted for a great deal. Like George Custer, who at Gettysburg wielded his saber to stab an oncoming rebel, a leader who developed a reputation for gallantry in battle usually found men willing to flock to his example. As one witness wrote of Custer's actions, "You can imagine how bravely soldiers fight for such a general."[21] That brand of "conspicuous gallantry" was understood to wield considerable power to motivate men in battle, and the Civil War was rife with examples of officers, particularly at high levels, embodying the ideal: the gravely wounded Daniel Sickles being evacuated from his line at Gettysburg, calmly puffing on a cigar, or Winfield Scott Hancock riding coolly among prone Union soldiers, apparently unruffled by the Confederate artillery barrage raging around him, steadying the Federal troops just before the onslaught of Pickett's Charge. At the same time, a more nuanced code of masculine behavior began to develop among many soldiers, who embraced a slightly more sophisticated understanding of the traits that constituted physical valor. Within these new conventions, a leader could demonstrate ability without a booming voice or showy personality by embodying a different kind of steadiness under fire. Aldace Walker's brigade commander Getty, who was in manner "very quiet," was nonetheless a "fighting man all over," as demonstrated by his apparent disregard for his own safety in pacing the skirmish line. The presence of such stoicism proved a powerful talisman in inspiring troops: when Walker's unit deployed in the summer of 1864, "The first question is always, 'Is Getty along?'" The reassuring answer was "invariably, 'Yes, on ahead.'"[22]

These examples of courage were most potent for the infantrymen who witnessed them directly. Walker wrote of another encounter with the diminutive Phil Sheridan, who despite his small stature had an electric effect on a disorganized group of Union soldiers. During an afternoon battle, Sheridan rode through the Federal skirmish line, shouting, "Crook and Averill are on their left and rear. We've got 'em, by G-d." Though Sheridan was not by reputation a profane man, his excitement was palpable, and, as Walker recalled, "it gave us all new courage." The effect of Sheridan's brief appearance was dramatic; Walker wrote that he and his comrades "weren't a bit demoralized" as before.[23] The tactics of the mid-nineteenth-century battlefield magnified the power of such techniques. Because the men were so close to one another and to their officers, it was easy for them to see and mimic the behaviors they observed around them. The comparatively modest-sized battlefields of the Civil War made it possible for even general officers to appear in person to significant portions of their commands, if only briefly, just as Daniel Morgan had circulated among his troops on the eve of the Cowpens battle. (Given the size of the battlefield and the magnitude of the armies engaged, such personal visits to large segments of a command in combat proved virtually impossible for World War II commanders.) The importance of modeling bravery had become, by the end of the war, an article of faith among military leaders. Indeed, two years after the end of the Civil War, the U.S. Army's 1867 tactical manual explicitly valued self-discipline and overt physical valor over protective cover, noting that "officers will observe that a too scrupulous regard for cover will make the men timid." Officers, the manual held, could influence the ranks surrounding them simply through their actions and "should therefore set the example of fearless exposure whenever an advantage can be gained."[24] The passage codified a belief that had taken firm root after the infantry's experiences in the past century and indicated the power that modeled behavior, whether to seek shelter from fire or to "fearlessly" brave it at every opportunity, was understood to wield over soldiers' actions.

At the same time, Civil War soldiers proved to be surprisingly harsh judges of men they pronounced unfit for command. The term "fragging" (the fratricidal practice of deliberately targeting one's own unpopular or incompetent officer, commonly by rolling a fragmentation grenade into an unsuspecting officer's tent) is associated most closely with the Vietnam War, though scattered evidence suggests that the same notion existed among more than a few earlier soldiers, who harbored homicidal feelings toward their officers even if they did not act on them. One New York volunteer warned his brother in the summer of 1863 that "some of the head officers here if they ever go in battle will not come out alive," and named his own adjutant as one example. He added that he had "heard a number of soldiers tell their own officers they would kill them if they ever get a chance."[25] Officers who did not stand up to the burdens of combat leadership warranted unforgiving treatment. One Confederate second lieutenant caught deserting found himself forced to run a gauntlet between two rows of men from his unit; as a witness reported, "any man was at liberty to kick his stern who felt like it." Privates, it seemed, enjoyed doling out the punishment most of all. In a final appropriate twist, the cowardly lieutenant was conscripted while packing to return home and, disgraced out of the officer corps, found himself carrying a rifle in the ranks at age forty-five.[26]

Enlisted men in both the Union and Confederate armies clearly chafed under standoffish or distant officers. Perhaps because of the nation's egalitarian traditions, the importance of the common touch became increasingly important to soldiers in the massive armies of the Civil War. Early in his army tenure, one young Union soldier evaluated nearby officers privately in his diary, singling out for insult those who acted better than the soldiers in their commands, neglected to return salutes from the headquarter guards, or walked with heads elevated so as to avoid the gaze of a private. Officers who merited praise included those who were not deluded by their own rank. One man, who "though he wears epaulets, can see little folk," warranted admiration; observing that there was "no stiffness about his salutes," the diarist pronounced him "the man to lead a company."[27] Such appreciation for the common touch in an officer was visible in the Continental Army as well: one eighteenth-century soldier offered praise for General Daniel Morgan at Cowpens after watching him mix easily with the rank and file, helping them affix their bayonets and joking with them about their sweethearts. Such easy familiarity cemented the private's respect for Morgan: "It was upon this occasion I was more perfectly convinced of Gen. Morgan's qualifications to command militia, than I had ever before been."[28] Evidence of the common touch increased both confidence and affection toward a combat leader. Similarly affectionate sentiments prompted Henry Welch of the 123rd New York Volunteers to write of his sorrow two weeks after losing a treasured colonel at the 1863 battle of Chancellorsville. "I miss Col. Norton very much," he wrote his aunt and uncle. "He was," Welch reflected, "a friend to me," proving particularly sympathetic during a bout of illness: "when I was sick with the fever he came often to see me and when I got better he was ever cautioning me to be more careful and not get down again." Norton refused to lord his officer's privileges over his men, and on the last march often went on foot while allowing Welch to ride. The colonel maintained an approachable demeanor ("He always spoke very pleasantly to the boys and he would as soon speak to a private as to an officer") and was, in Welch's judgment, "one of the finest men that I ever saw. He was in fact one of nature's noblemen."[29]

Reference to an officer as one of "nature's noblemen" encompassed a powerful implicit logic. Combat leadership, it suggested, was an inborn ability. Like their predecessors in the War of Independence, most nineteenth-century American soldiers assumed that the combat leader's flinty ability to stand firm in the face of hostile fire was essentially an innate characteristic, native to a certain breed of warrior. A *Harper's Weekly* article that appeared during the Civil War reiterated a popular belief that "war being an art, not a science, a man can no more be made a first-class painter, or a great poet, by professors and textbooks; he must be born with the genius of war in his breast."[30] Above all, infantry combat in the nineteenth century demanded leaders who could maintain an unruffled appearance amid the danger and chaos of the battlefield, a staggeringly difficult but relatively straightforward task, and one that seemed to depend more on intrinsic qualities of character than on any sort of technical competence.

Not surprisingly given the extent of tactical and technological changes on the battlefield, attitudes about both the nature and manifestation of leadership shifted dramatically in the twentieth century. Certainly, the dispersed tactical system did not lessen the men's need for active direction in combat. That need continued as strong as ever: as a World War II veteran recalled, "Regardless of the situation, men turn to the officer for leadership, and if he doesn't give it to them then they look to the strongest personality who steps forward and becomes a leader," whether that man happened to be a staff sergeant or a platoon leader.[31] But the nature of that leadership evolved significantly as the twentieth century's empty battlefield placed considerable new demands upon small-unit leaders. As infantry combat became more spread out, it became far more difficult for a lieutenant to observe his soldiers' actions firsthand; dispersion, combined with changing societal expectations, made it far more difficult to employ coercive threats to get soldiers to brave enemy bullets. The dispersed battlefield thus robbed both components of the eighteenth- and nineteenth-century officer's highly effective carrot-and-stick tandem of much of their motivational power. That was the essence of an acknowledgment in a 1917 officer's manual, which held that "the discipline that will insure obedience under any and all conditions . . . to march, to attack, to charge" was equally important in the dispersed system, even though it could not be attained "by the machine-making methods of former times." The necessary difference in leadership style was acute: "The company commander used to drive; now he leads."[32] The shift from a system that demanded soldiers to behave as automatons to one that required them to exercise considerable autonomy triggered similar changes in the nature of leadership provided them in the thick of battle. At the same time, the increasingly deadly character of twentieth-century combat heavily taxed the army's system for producing officers. The enormous armies of the mid-twentieth century demanded a substantially larger group of officers to lead them, and the casualty rates and continuous character of modern infantry combat led to correspondingly higher turnover in personnel, requiring a larger pool of officers from which to draw replacements. Samuel Stouffer's team made the calculations with brutal simplicity: though second lieutenants constituted slightly less than one percent of a division's total strength, they accounted for 2.7 percent of all battle casualties. With an average daily casualty rate of 11.4 per 1,000 men in the European Theater, the outcome of a lengthy period in a forward area was both stark and chilling: "Assuming, for the sake of demonstration, that the full complement of 132 Infantry second lieutenants was present in a division on each day of combat," the Stouffer team estimated "the division would lose a full complement of its Infantry second lieutenants in 88 combat days."[33]

The new burdens of high-speed, modern warfare helped usher in the twentieth century's more organized, analytic assessment of combat leadership. Confronted with much higher demands for manpower, the army could no longer afford the belief that the ability to lead troops in combat was simply an innate quality. Nor could it depend upon a system devised simply to identify and cultivate leadership ability in those who had already possessed it innately. Psychologists and sociologists, armed with survey and experimental data, argued that effective combat leadership consisted of a set of skills that could be studied, developed, and, most importantly, taught and learned. The American army, faced with the need to obtain more and more officers and place them in a tactical system that tended to discount many of the personal characteristics upon which earlier combat leaders had depended to motivate troops, was more than willing to be persuaded. Thus by 1943 the U.S. Army had given the first chapter of its book *Leadership for American Army Leaders* the title "Leadership Can Be Learned," and trumpeted confidently that "the theory that leaders are born and not made is the saw of the defeatist."[34]

Higher rates of attrition in battle reinforced the importance of formal organization over the personal ties that shaped leadership on the battlefields of the eighteenth and nineteenth centuries. Stouffer's team found that, in infantry outfits, "the tremendous turnover of both men and their officers during combat minimized the possible importance of strong personal attachments at the level of company or smaller units." Because second lieutenants suffered casualties as high and usually higher than the riflemen they led, "it was not unusual to find a rifle company which after two or three months in combat had none of its original officers remaining in the unit."³⁵ That high personnel turnover dictated that the system replace personality, since so many officers were being killed so quickly. The pamphlet *Psychology for the Fighting Man* spoke directly to this problem: leaders in an army, it argued, must be "interchangeable." When one small-unit leader was killed, "another must be ready to take his place and lead his men." Loyalty to the system, rather than to a specific personality, ran both ways: "the men who lose a leader must be ready to follow without question the commands of a stranger."³⁶

As they gained combat experience, GIs of the Second World War developed a heightened appreciation for the increased demands under which their officers labored. Most obvious was the fact that leaders had tactical responsibilities that often precluded taking cover on the empty battlefield. An infantryman of the 99th Division observed during his service in Europe that officers "had so much responsibility and couldn't keep under cover so the attrition rate was high."³⁷ The nature and magnitude of the risks officers ran was not lost on the men they led. Offered a slot in Officer Candidate School (OCS), one World War II GI reported that he did not consider the opportunity very seriously. "Having witnessed the high mortality rate of junior officers during the Battle of the Bulge," he wrote later, "I declined the offer without reservation." Though a comrade attempted to emphasize the "great opportunity" that OCS represented, his decision remained firm. Combat experience bore out the wisdom of his choice: by war's end, nineteen of his regiment's junior officers had been killed, nearly all of them platoon leaders.³⁸ (While experienced soldiers often came to appreciate the increased physical dangers their officers endured, the intangible burdens of leadership proved more difficult to appreciate. Particularly taxing was the necessity of sending men to their deaths, and the guilt that attended that responsibility. In November 1944, Thomas Bishop, a lieutenant colonel in the 99th Division, recorded in his diary that one of his soldiers had been "reported killed at 1200 while on patrol." Bishop sent out another patrol to retrieve the body, expecting to "thus lose additional men, I am sure."³⁹ The unrelenting pressure of performing the grisly calculus of men's lives weighed heavily on many combat leaders, a burden that few enlisted men on the line seemed to comprehend.)

Attitudes among the rank and file toward physical bravery in their officers mirrored the changes in attitudes about the nature and meaning of fear outlined in chapter 2. Soldiers still treated shows of physical bravery with respect or disbelief: decades after the Second World

War many veterans still wrote with awe at the memory of a particular officer or NCO rushing through a hail of gunfire past men frozen in their tracks. But realism tempered that appreciation. Leaders who took unnecessary chances were more likely to be branded reckless, or worse. The "fearless exposure" so important in leaders on the nineteenth-century battlefield was usually suicidal on the technology-dominated modern battlefield; repeated exposure to automatic weapons and high-powered artillery, no matter how fearless, quickly proved lethal. Instead, competence became the standard measure for an officer's ability in the eyes of most men. In a chapter entitled "What Soldiers Think of Leaders," the *Psychology for the Fighting Man* pamphlet cataloged a dozen traits that enlisted men associated with good leadership: "competence" came first, followed by "interest in the welfare of the soldier," "promptness in making decisions," teaching ability, and "common sense." Sense of humor, physique, and impartiality were among the traits that rounded out the list; conspicuous displays of fearlessness did not appear at all.[40] That is not to suggest that soldiers had discounted the importance of courage in their leaders. Valor still held enormous importance, but expectations of its expressions had changed. Nearly 60 percent of Stouffer's enlisted interviewees listed "courage and aggressiveness" as the most desirable characteristic in a combat leader. But soldiers no longer evaluated courage as simply as had their predecessors in the armies of the eighteenth and nineteenth centuries. An officer who trembled (or even cried occasionally) during a firefight could still win the esteem and respect of his troops as long as those outward signs of fear did not compromise his combat performance. Because the empty battlefield placed relatively less importance on modeling stoic behavior, interpretations of courageous behavior assumed new flexibility. Most soldiers had seen that such involuntary physical tics indicated little of a leader's ability to hold up to the rigors of battle. In *How to Get Along in the Army*, "Old Sarge" directly cautioned soldiers against judging combat ability by such outward appearances through a description of the strongest soldier in his unit, a man who nevertheless fainted in the presence of needles. Sarge praised the man's strength ("He could have broken me in two—and I'm no weakling"), yet the mere sight of a syringe dropped him to the floor. It was, Sarge suggested, "something psychological," and he cautioned soldiers not to hold themselves to an unforgiving standard: "don't be ashamed if you're in his class."[41] Combat experience also demonstrated that a tough outward appearance could be a misleading facade: sometimes, the toughest, most physically imposing soldiers turned out to be the weakest in battle. Such was the experience of a private during the Battle of the Bulge. A burly platoon sergeant from another division (who, as it happened, had made life miserable for many of the men in training) "cowered in his hole" during the battle itself. German soldiers laughed at the sergeant's pathetic performance before shooting him.[42] Twentieth-century soldiers who witnessed such dissonance between appearance and performance (and there were many) recognized a useful pattern.

The nature of infantry combat in the twentieth century's dispersed tactical system helped prioritize combat ability over the straightforward mastery of one's outward physical appearance. Effective decision-making replaced unflinching exposure and urging on the soldiers of the company as the officer's critical task on the battlefield, because it was effective decision making that both achieved military goals and kept individual soldiers alive. New emphasis thus fell upon training leaders whose judgment remained unaffected by the maelstrom of combat, whatever their outward appearance might be. Training leaders for the empty battlefield included the same attempts to introduce realistic elements that also featured in World War II training regimens for enlisted men. Such was the logic behind one OCS exercise that attempted to discern a candidate's ability to deal with pressure by placing him in two "stress exercises." Working against a timer, the subject attempted to solve a map problem and to translate codes while he listened to piercing battlefield noises pumped through an earphone, suffered periodic shocks administered through a wrist device, endured the sporadic vibration of his chair, and encountered difficulty breathing owing to the partially closed gas mask he wore: a realistic test that clearly prized results over calm appearances.[43]

The influence of tactics and technology on modern infantry combat triggered another important shift in the average soldier's experience of combat and his attitudes toward leadership. Effectiveness in nineteenth-century combat depended on the literal cohesion among infantrymen on the field: the noise, confusion, and crude communications systems of battle meant that formations had to remain intact if they were to maneuver and deliver fire effectively. Holes in the line, whether due to wounded men dropping out or panicked men fleeing, compromised the unit's solidarity. Too many gaps in the line rendered the remainder of the unit precariously vulnerable. A New York officer at Fredericksburg in 1862 who witnessed the devastating Confederate fire on Union formations described its effect upon the troops' combat effectiveness, noting that "when the ranks are torn by artillery, the cohesion begins to fail." The situation devolved precipitously from there: "Then expose the men for several hundred yards to a murderous fire of musketry, and front man rank is gone, rear man rank is gone, comrades in battle are gone too." Ultimately a formation lost its integrity completely, as "a few men struggle along together, but the whole mass has become diluent." The remainder was but a pale remnant of the unit's original strength, and the officers could do little with them: "Little streams of men pour in various directions. They no longer are amenable to command."[44] Those pragmatic observations, exceedingly common on the linear battlefields of the eighteenth and nineteenth centuries, explained soldiers' appreciation for stoicism in the face of enemy fire (some of the loudest cheers on the battlefield at Waterloo, for example, went up for a unit that withstood a withering enemy barrage, and took heavy casualties, without flinching). Just as individual soldiers depended on their neighbors for protection and support in the line, a company in battle depended on its neighboring companies to shield its flanks. These tactical and technical realities of eighteenth- and nineteenth-century combat also explained why independent action was such a small part of battlefield leadership at the junior level. On the linear battlefield, a captain who decided to evaluate the tactical situation independently and led his men out of formation to exploit a perceived opportunity invited disaster, not only for his company but often for neighboring units as well.

Combat within the modern system effectively inverted this construction, making initiative in a leader not just a prized trait but an absolutely indispensable one. On the empty battlefield, combat moved far more quickly, and engaged units had to respond to events much more fluidly and rapidly, making decisions with imperfect information. In the presence of automatic weapons, stoically enduring fire in the open would reduce a platoon to tatters in moments. Cover, concealment, and coordinated movement provided soldiers the only possible protection by making them more difficult to strike. GIs on the battlefield grasped those truths quickly and came to demand action—any kind of action—from their leaders. Samuel Stouffer's team of social scientists interviewed one wounded veteran whose gripes about his commanding officers revolved around their refusal to act: "One time we begged our lieutenants to give orders" in the middle of a firefight. Those lieutenants "were afraid to act because they didn't have the rank": that is, they worried that some nearby captain or major remained in command. To that veteran, the unit's ordeal stemmed directly from the lieutenants' refusal to seize the initiative. "We took a beating while they were waiting for orders," he indicated, wondering how the lieutenants knew their commander "hadn't been knocked off" in the meantime.[45] Active participation in battle helped reinforce the fragile notion that at least some parts of the combat experience remained controllable. And action, particularly the kind of focused action provided by competent leadership, offered some means for soldiers to strike back at the threats directed against them—a useful, if sometimes illusory way to encourage soldiers in the belief that they could improve their own chances for survival. Physical activity also worked to counteract the stresses common to soldiers under fire and countered the sense of helplessness and passivity that proved so detrimental to overall combat effectiveness. Psychologists frequently prescribed work, even "chickenshit" busywork, to help prevent idleness and depression. Those lessons appeared in the army's own morale materials, reminding readers that "not only does morale make soldiers work and fight. Working and fighting keep up their morale."[46] Physical outlets for anxiety carried a perceived utility, even when the activity itself was of little objective value: one Marine captain forced to take cover in the Pacific wished he could ease the strain of Japanese air attacks by firing his pistol at the planes or even by hurling sticks at them.[47]

The more clinical understanding of the symptoms of fear in the twentieth century, as products of natural physiological responses rather than as moral failings, did not completely overwrite traditional conceptions. Rather, because split-second battlefield decision making exerted such a pronounced influence over a unit's survival in combat, World War II GIs tended to respect leaders who maintained the ability to think and function amid the chaos of combat, and usually forgave the occasional physical outburst so long as it did not seem to cloud judgment. Accordingly, sections of the Army's 1941 Field Manual 100-5 dealing with leadership and command called for "cool and thoughtful leaders with a strong feeling of the great responsibility imposed upon them." "A leader," the pamphlet argued, "must have superior knowledge, willpower, self-confidence, initiative and disregard of self."[48]

The final, and in many ways most important, criteria for soldiers assessing a leader's abilities during World War II was the degree of responsibility the leader demonstrated with the lives of the men under his command. For GIs of the Second World War, the emphasis lay unequivocally (and not unreasonably, given the ordinary soldier's overriding desire to stay alive) on thrift: troops demanded officers who minimized the threats to which they were exposed, prizing those who got them through battle with a minimum of risk and, by extension, minimal casualties. Soldiers therefore often judged the most effective officer at the company level to be the one who fought alongside the men he led and did not, in the words of one 78th Infantry Division GI, "needlessly sacrifice the lives of those under their command."[49]

Such pragmatic assessments of leadership created some insoluble tensions for American officers in World War II, as soldiers' definition of "needless sacrifice" frequently clashed dramatically with what headquarters demanded. Men often judged a leader by his success in keeping them alive, while his superiors judged him by his success in achieving military objectives, twin goals that often proved mutually exclusive. That tension between expectations from above and from below helped account for some of the enlisted men's harsh appraisals of their officers' leadership acumen. Officers' struggles with their own fears accounted for others. In evaluating the junior officers he saw in the ETO, one GI praised his own lieutenant while criticizing a number of others: some were cowards, some were "simply trying to stay alive and did a lot of CYA" (army slang for "cover your ass") rather than leading troops.[50] At the Battle of the Bulge, private Frank Raila remembered losing respect for his company commander when he witnessed behavior that, to him, constituted shirking responsibilities: "I had contempt for the captain of my company when I saw he dug his foxhole next to the one of the captains from another company." Both men, in Raila's estimation, "appeared to lack leadership," presumably because their chumminess interfered with their management of the men entrusted to their command. The fact that the two captains stuck together while the enlisted men milled around without direction irritated the private. "It seemed to me," Raila wrote later, "they should have been out directing us, advising us, encouraging us."[51] Particularly singled out for criticism were newly minted officers, whose ability to empathize with their charges was usually found wanting by more experienced soldiers. "New officers who come over here," claimed a veteran of the Italian campaign, "can't understand a man's feelings after living the way he does over here."[52] Experienced soldiers also viewed new leaders' technical competence with suspicion—often with good reason. As noted in earlier chapters, modern industrialized battle presented combatants with a dizzying array of weapons, technologies, tactics, and techniques to master. The learning curve appeared frightfully steep; as many veterans later noted, once in battle they quickly discovered that actual combat bore little resemblance to the exercises described in training manuals, and many of the soldier's most important techniques for fighting and staying alive were only acquired through hard-won battle experience. A skilled veteran, for example, could often determine whether a nearby weapon was American- or German-made simply by the distinctive crack of its report, and could establish swiftly whether an artillery round was incoming or outgoing by the tone of the *whoosh* it produced. Men looked up to leaders who had mastered the techniques of soldiering: officers who could read a piece of terrain and locate the dead spots quickly or who knew the effective technique to flank a machine-gun nest. Rank-and-file soldiers attached high value to their leaders' battlefield acumen for a simple reason: those leaders' technical know-how increased their own odds of survival. John Dollard's American veterans of the Spanish Civil War voiced a strong preference for experienced soldiers: "Practically all our informants agree that going into action with a tested man made them better soldiers." Those soldiers offered two reasons: an experienced leader understood "how to accomplish objectives with a minimum of risk" and, in his demeanor, set "an example of coolness and efficiency which impels similar behavior in others."[53] Both notions appeared over and over in official booklets. The first qualities men appreciated in a leader, according to the official line in *Psychology for the Fighting Man*, were "competence and ability." "Competence," the pamphlet argued, "is based on learning." An effective leader, it held, knew not just the things learned in training courses but also kept up to date on new techniques and tactics. The rule, the pamphlet counseled, was "a simple one: *know your stuff*."[54] Those lessons were supremely difficult to teach in training, and the army could rarely spare its experienced officers to instruct candidates. As the war dragged on, the perilously high turnover rate thus frustrated many new officers' attempts to win the respect of their charges, who (especially when those new leaders first took command) knew more about the business of infantry combat than their untested leaders. New lieutenants often found it difficult to survive long enough to accumulate the combat experience that would convince the men of their ability. One World War II infantryman reported particular frustration with West Pointers, complaining after the war that the service academy sent second lieutenants to the front, where they lasted only two or three weeks on the misleadingly named "empty battlefield" simply "because they just didn't believe the enemy was there."[55] Other soldiers were exasperated by supposedly educated professional soldiers whose classroom training helped little in battle and whose lessons, it seemed, often eclipsed common sense. Years after the war a GI from the 36th Division groused about an unpopular colonel whose four years of classroom instruction had little utility in the field. The colonel's inexperience was particularly frustrating to his veteran troops, contributing to tension within the unit on one occasion in particular as the colonel looked over a diagram of their position: "He probably had four years of West Point instruction in tactics concentrated on that poor little map, with its lines and cross lines and arrows and gun positions and concentration areas and routes of withdrawal, all prettied up in red and blue grease pencil." The men were unmoved by the display of book learning and sat there "staring at him through eyes bloodshot with exhaustion yet somehow fascinated with the whole foolish, frightening scene." The colonel's exaggerated theatrics ("Suddenly, he leaped to his feet, awkwardly drew his pistol and shouted excitedly, 'All right, men, on your feet, they're attacking us but we'll defend this command post to the death!'") only emphasized his preposterous figure: "All we could hear was the distant rattle of machine-gun fire. 'Colonel,' someone said pityingly, 'that's a mile or more away.'"[56]

Such dismissive attitudes, and the unforgiving standards of competence they represented, grew logically out of a tactical system that placed the burden of immediate decision making in combat squarely on the shoulders of small-unit leaders. On the linear battlefields of the War of Independence and Civil War, a unit's fortunes in battle rested in large part on factors far beyond the control of the company's captain. There, a regiment's position in the line was usually determined by a colonel or general, and once battle began comparably few tactical decisions occurred below that rank. Companies, and the regiments they composed, took their positions and blazed away mechanically as long as they could manage; their captains and lieutenants stood nearby, offering encouragement (and the occasional threat) to keep soldiers focused amid the barrage. Combat on the empty battlefields of the mid-twentieth century placed a radically different set of demands on lieutenants, captains, and their noncommissioned officers. A platoon leader might make dozens of split-second decisions during the short span of a firefight that determined whether his unit carried out a mission without a casualty or whether enemy fire cut the troops to ribbons. To an even higher degree than on the linear battlefields of the eighteenth and nineteenth centuries, an incompetent or ineffective leader on the dispersed battlefield posed a serious threat to his troops' survival.

The image of the out-of-touch leader whose office insulated him from the reality of the privates' war persisted throughout the conflict. The image could be by turns tragic and comic: tragic when it led to the senseless loss of men's lives by foolhardy or inexperienced officers; comic when it served to reinforce the men's sense that they, in fact, knew the business of soldiering better than their officers. A Bill Mauldin cartoon, depicting an exchange between a grizzled infantryman and a new lieutenant who had just lit a cigarette during a nighttime encampment (a forbidden action that threatened to draw enemy fire), illustrated this gulf with a wink to enlisted GIs. Mauldin showed the haughty, smooth-faced lieutenant chastising the unshaven private ("You're not on the job, sentry — a sniper almost got me") as the sentry's exasperated expression and smoking M1 leave the audience little question as to where the offending bullet originated.[57] Conversely, World War II soldiers treasured leaders whose easy manner bridged the gulf between officer and enlisted man, as had their counterparts in eighteenth- and nineteenth-century American armies. Pyle described one colonel who clearly enjoyed the affection and respect of his men despite his schoolteacher appearance. The colonel "cussed a blue streak" and called the soldiers by their first names to lessen the gap between his rank and the men. The same colonel eschewed the official insignia on his cap, as well, a useful practice in two senses: it kept him from being singled out for special treatment both from his men and from enemy snipers.[58] A combat engineer from the 1st Infantry Division reserved special praise for a first-rate officer who distinguished himself from OCS officers by his solicitous manner: "He would get us together. He said, 'Now men they told me that we've got to go out and secure this area.'" The officer asked the troops' opinion of how many soldiers he should take, and asked for a volunteer to accompany him on the mission. The engineer commended the officer as "dead honest. He was a man you knew was a good leader because he would ask you for your opinion."[59] The nature of combat in the dispersed system made such conduct more than simple social pleasantry. It sent the message that the officer was willing to avail himself of the combined wisdom of the unit's soldiers in planning a mission, provided further evidence of his concern for their survival, and further reinforced soldiers' sense of decision-making power (and, by extension, their personal agency) by actively seeking their input.

And of course men appreciated (and fairly demanded) officers who demonstrated concern for their well-being and survival, along with a willingness to share in the unit's risks. Ernie Pyle wrote of one such officer, whose attention to small details (refusing a drink of liquor, for example, unless in possession of enough for the entire unit) helped bond the men of the outfit together. Not all the officers Pyle encountered were as selfless, but such actions helped bond small units together. "All officers are not like that," he wrote, but the "common bond of death draws human beings toward one another over the artificial barrier of rank."[60] When officers failed to bridge the gap between the rank and the men, the tensions often crystallized around class. Those divisions became particularly sharp in the Second World War, with a bevy of privileges (better food, officers' clubs, and other perks) accruing to officers. One of Mauldin's most famous cartoons, in which an officer admiring a magnificent sunset in the mountains wonders if a comparable view has been furnished for the enlisted men, evoked the particular pretension so many men detected in their officers. Certainly those class battles were not limited to the dispersed battlefields of the twentieth century; combat leaders on both sides of the watershed encountered some similar tensions. In his work on the changing system of values in the Civil War, historian Gerald Linderman argued that many nineteenth-century soldiers viewed West Pointers as "class-ridden snobs" who traveled "in the fashionable circles of metropolitan society."[61] After the Civil War, General William T. Sherman attributed many West Pointers' poor ability in handling men to something akin to arrogance: as he told a congressional committee, few of those who graduated near the top of their classes became distinguished military leaders. "They are taken," Sherman argued, "from the head of their classes at West Point. They are good scholars, but . . . begin to look down upon the rest of the army."[62] In other regards, however, the expectations of eighteenth- and nineteenth-century soldiers appeared far more forgiving than those of World War II GIs. Infantrymen of the War of Independence and the Civil War harbored little hope that their officers could keep abreast of the war's overall progress: the size of the undertaking, the limited reach of news and the slow speed with which it traveled, and the already overburdened logistical systems meant that nearly everyone remained cut off from news. A soldier in the Army of the Potomac in July 1863 was as likely to learn of Grant's capture of Vicksburg from his parents as from the captain of his company. Soldiers of the mid-twentieth century held their leaders to a different standard in this respect. Just as the twentieth-century infantryman wanted to know more about the whys of his training (not just how to squeeze a trigger, for example, but why it should be squeezed rather than pulled), those soldiers also wanted to understand the importance of the risks they would face in battle. They expected their officers to inform them how their sacrifices fit into the war's overall picture. Jim Bowers noted that this expectation began in basic training: "One of the features about the field lectures I enjoyed was that a noncom ended each day's lecture with a brief telling of the day's war news."[63] Failure to keep the men abreast of the war's progress once the soldiers deployed in theater engendered bitterness between leaders and those who served beneath them; in addition, while it might not call the leader's competence into question, it often drove a serious wedge between a soldier and his officers. "I thought they knew what they were doing," one GI said of some noncommunicative officers, "but they did not inform us of anything." In this case, the officers didn't tell the men where, specifically, they were or who the unit was replacing. Enlisted men often cast the frustration that the lack of information engendered in class terms: such tight-lipped officers "didn't seem to believe in communicating with the lowly dogfaces."[64] Left unchecked, social scientists predicted, such cracks in a unit's cohesive bonds would eventually have dire consequences on the battlefield. One infantryman recalled that "the leaders did not let us know what was happening or what we were going to do." The officers' apparent "ignorance bothered me very much. The leaders did not make me feel like part of the group."[65]

Nor could leaders entirely relax their attention to outward appearances. While the increasingly universal understanding of fear symptoms as a naturally occurring part of the body's response to combat allowed leaders a more realistic range of responses in combat, the men they led still took important cues from their officers' appearance and presentation. The second chapter of *Psychology for the Fighting Man*, "The Manner of the Leader," instructed officers to consider how their appearances affected their men, who looked to a leader's demeanor and tone for hints about what went on in the officer's mind. Recalling the stoicism so prized on the battlefields of the eighteenth and nineteenth centuries, the pamphlet deemed that "the cultivation by leaders, and particularly officer-leaders, of a calm, controlled manner is essential. The leader needs mastery over facial expression, control over voice and gesture."[66] Equally critical to leadership in the Second World War was the use of language; it became a critical if sometimes subtle tool for officers to influence soldiers' perceptions and actions. In instructing combat leaders to give thought to the subconscious cues displayed in an officer's bearing, the *Leadership for Army Leaders* pamphlet paid special attention to language and its influence. Counseling officers to avoid histrionic bluster, the pamphlet noted that such outbursts suggested a lack of control on the part of the officer. The effect could be ruinous: "Such lack of control lowers a leader's prestige and may even make him laughable. Donald Duck is funny, and so may any man be who commonly bursts into tantrums. In sum, the manner of the leader toward his men makes for or against cooperation."[67] One of the most useful ways to employ language in a combat situation was to use it as a tool that emphasized soldiers' own sense of agency and their perception of the battlefield environment as controllable to some degree. Many of the most effective officers employed language in this fashion almost instinctively, naturally choosing phrases that emphasized the infantryman's sense of control. *Psychology for the Fighting Man* dedicated an entire chapter, "Language of the Leader," to the ways in which officers could build this sense of efficacy through well-chosen words. Recommending "short, uninvolved, and incisive" sentences, the pamphlet counseled leaders to avoid vague, ineffectual, or indecisive words: the officer's language "should be positive and direct, not uncertain, inconclusive, or negative." Specific cases illustrated ways for officers to appeal to the men's agency. Phrases like "I'm sure you can do it," "You're just the man to do it," "There must be a way—I know you can find it" would produce "confidence, self-reliance, and determination" among the rank and file. The converse was true, as well: phrases like "Maybe you can do it," "See if you can't do it," "I doubt if you can do it—but go ahead and try" would bring "doubt and wavering."[68] Soldiers' own combat experience testified to the effectiveness of that kind of language. As his unit struggled in battle along Elsenborn Ridge, infantryman John Campbell remembered a brief visit from a brigadier general. Asked about the progress of the battle, the general replied in vigorous and determined terms. Campbell could not recall the general's precise language (fifty years later he had forgotten whether the general said "Hold out boys, we've got them in a couple of days" or "We're going to kill 'em!"), but the positive expectations revealed in the tone of his response stayed with Campbell for years and provided an enormous boost for the men on the line: "It was the best thing we could have heard."[69]

One of the most common and specific uses of language by officers to motivate troops, of course, came in the form of the "pep talk" delivered before battle. Such spoken encouragement was hardly new to the twentieth century; early American military history offers scores of examples of officers addressing troops before an engagement in an effort to bolster their motivation for combat. Daniel Morgan's encouragement to the militia on the eve of the battle at Cowpens provided one

excellent example of effective inspiration. At base, Morgan's instructions were quite reasonable. Rather than ask his militiamen to hold a
position at all costs, he gave them a predetermined and manageable
task: "Just hold up your heads, boys, three fires . . . and you are free."
In a remark evocative of the St. Crispin's Day speech in Shakespeare's
Henry V, Morgan projected his men into a successful future, just as
Prince Hal imagined his soldiers celebrating the anniversary of their
presumed victory: "And then when you return to your homes, how
the old folks will bless you, and the girls will kiss you, for your gallant
conduct!"[70] The language itself was significant; references not to "if"
but "when" the militia made its safe return home encouraged soldiers
to embrace their duties in the coming battle as both eminently feasible
and destined for success. A Civil War colonel delivered a similarly terse
address as his unit prepared for action at Antietam. In that instance
the officer appealed to both the men's sense of honor ("You have never
disgraced your State; I hope you won't this time") and to their sense of
self-interest. "If any man runs," the colonel assured them, "I want the
file closers to shoot him; if they don't, I shall myself. That's all I have
to say." It was, in one soldier's estimation, pitch-perfect: "it's what I
call a model speech."[71]

Leaders' use of language to reinforce notions of individual agency
became even more important on the dispersed battlefields of the Second World War. In combat there the pep talk assumed a heightened
importance, since an officer might not maintain close contact with
the men during the actual fighting. A World War II GI remembered a
particularly effective reminder, one that was "simple and sweet, short
and to the point"; the officer's message became "etched in the granite
of our minds." Noting that the following day might be the most important of the soldiers' lives (and that "for some of you it will be your
last"), he urged each soldier to make the most of it. He counseled them
on several principles that would both increase their individual chances
of survival and would improve the unit's pursuit of its goals: "Put your
training to good use—don't bunch up, stay as close to the enemy as
possible, use marching fire, keep going so the medics can handle the
wounded—remember your basics."[72]

Other pep talks fell along a spectrum that ranged from over-the-
top masculine bravado to more restrained, contemplative discussion of
the trials ahead. Both types of address could coexist within the same
unit. On the night of June 5, 1944, American airborne infantry units
received a variety of encouragements from their leaders as they prepared to jump in support of the Allied landings at Normandy. The

nature of these talks varied dramatically depending on the regiment. In the 501st Parachute Infantry Regiment, Carl Cartledge and his fellow troopers heard a speech from their regimental commander in an atmosphere electric with the excitement of impending battle: the "juice was running in us," Cartledge recalled years later. The colonel "gave a great battle speech, saying 'victory' and 'liberation' and giving death to the enemy, and some of us would die, and so on." Forming the regiment into a line, he proceeded to shake the hand of each soldier, and then, in a final melodramatic touch, "he reached down and pulled his knife from his boot and raised it high above his head, promising us in a battle cry, 'Before the dawn of another day, I'll sink this knife into the heart of the foulest bastard in Nazi-land.'" The gesture connected dramatically with the men, as a "resounding yell burst forth from all two thousand of us as we raised our knives in response."[73] Further down the chain of command in another unit, a sergeant in the 508th Parachute Infantry Regiment received a markedly different pep talk just two hours later. As his company formed up to depart for their planes, his company commander talked to the men and presented what the sergeant described as a "well-thought-out speech." They were prepared, the captain said, and in a nod to their role as shapers of their own fortunes, "it was up to us to use that preparation and ingenuity to take care of the Germans." Like Cartledge's colonel, the captain's parting words dealt with death in combat, but took a markedly different tack: "He then said that he would rather die than have to bury one of his men."[74] Effective use of language could look both ahead to the future and backward to tradition. One new captain addressing his men bundled expectation and history together into a potent combination, reassuring his men of his pride in the unit and informing them of his expectation for their achievements in battle: "Men, I want you to know that I am extremely proud to be leading you into combat and that I will be proud of your performance." Without sidestepping the reality of their situation ("We will be given tough assignments and will take casualties"), the officer maintained firm expectations. "Many, perhaps most, of us will be wounded," he concluded. "Some will die. We will be afraid, but will not shirk. This is what our country expects of us—to serve as well as our fathers and grandfathers did."[75] The appeals to the men's sense of honor, and the reference to their place in a long lineage of American soldiers, constituted a potent challenge to their own masculinity. Soldiers' responses to these pep talks were as varied as the speeches themselves, from enthusiasm to cynicism. One of Bill Mauldin's cartoons captured a pragmatic incarnation of the latter emotion perfectly, as men in a trench asked the one-star general standing before them, "Sir, do ya hafta draw fire while yer inspirin' us?"[76] Such sarcasm underlined an equally important point: for many veteran soldiers, the rhetoric of the prebattle exhortation could never eclipse the grim realities of combat.

Strong leadership proved critical to channeling troops' focus in combat and preventing them from fleeing in these three conflicts time and time again. Perhaps no trope is quite so common in veteran soldiers' memoirs and war stories as the anecdote of a determined leader whose singular actions made the critical difference at a vital juncture in a battle. But, as soldiers themselves came to appreciate, the presence of an effective leader alone did not dictate the outcome of a particular firefight. In his diary, a Union soldier described one such situation, an 1862 action between his unit and members of the famed Confederate Stonewall Brigade. The contest had been "a fair, open fight, decided by pluck and discipline"; the soldier noted that the Union advantage in numbers balanced the Confederates' superior tactical position. The members of the Stonewall Brigade, however, had the benefit of "a leader in whom they had implicit confidence," while the Union infantry had "no leader but were commanded by a senior colonel" who remained more or less anonymous. Despite the advantage of a well-known and trusted leader, the rebels were "soundly thrashed"—due, in the diarist's judgment, to the "superior fire and courage" of the Federal infantrymen, regardless of whatever perceived deficit of charismatic leadership they suffered. The comment that the outcome of the battle made upon the importance and the limitations of leadership in battle did not escape his attention; the diarist closed the entry with a smug, "Thus endeth the first lesson for this valley."[77]

As that Civil War soldier and others like him learned, no individual could alone dictate the outcome of a battle or a skirmish. In both linear and dispersed systems, the leader's most important duty was to orchestrate the way troops employed their weapons in battle: forming individual soldiers into a unit that could unleash its destructive power in concert to maximize group effort. Since the beginning of the gunpowder age, portable firearms have provided infantry soldiers with a way to channel their wills into military power that could be directed at an enemy. The way individual behavior in combat changed as foot soldiers came to possess ever more accurate and quick-firing weapons, and soldiers' evolving relationships with that weaponry, provides the subject for the next chapter.

5

Weaponry

Robert Walker formed part of the first waves of American infantry-men who assaulted the beaches in northern France the morning of D-Day in June 1944. Walker's arrival on the continent was somewhat inauspicious; his landing craft capsized during the journey to shore, and he lost his rifle in the confusion. Waterlogged and weaponless, he waded to the beach. Despite all his training had done to solidify his faith in his skills as a warrior, Walker had difficulty even picturing himself as a soldier: "Instead of being a fierce, well-trained, fighting infantryman," he recalled, "I was an exhausted, almost helpless un-armed survivor of a shipwreck." American paratrooper Tommy Horne was also part of the massive Allied invasion, dropped into France hours before the armada crossed the channel to help prepare the way for the infantry arriving at daybreak. In the confusion of the jump, Horne, too, lost his rifle. Landing in a field covered with dead bodies, he scavenged and discarded two rifles with deformed barrels before locating a usable weapon: "I finally found a rifle that wasn't bent or had a broken stock, and I got two bandoliers of ammunition off of one of the dead fellows." Newly rearmed, he remembered, the battlefield he surveyed suddenly became more manageable to him: "I felt pretty good. I had a weapon."[1]

The difference in the two soldiers' responses is instructive. Human elements—training, leadership, discipline, willpower—constituted critical factors that shaped the ways battles unfolded. As earlier chapters suggested, the internal struggle, mastering fear long enough to stay and fight, affected soldiers no matter when or where they fought. But merely enduring the trauma of combat was hardly sufficient by itself; to be effective, soldiers needed to act in constructive ways. Pairing soldiers with weapons was instrumental, since those weapons constituted the tools with which to respond to the various threats of combat. It is of little surprise, then, that individual soldiers' willingness to fight frequently was linked so closely to the weaponry itself. One paratrooper's comment on the unit's tense preparation revealed his persistent belief that his actions would bolster his chances in combat: "Our lives depended on our weapons functioning properly when needed — and on our ability and willingness to use them."[2]

Fear, and the desperate desire to stay alive, connected a soldier's arms with the willingness to stay and fight. Foot soldiers of the eighteenth, nineteenth, and twentieth centuries experienced terror in no small part because they were so vulnerable to enemy fire. By balancing their feelings of vulnerability with the perception that they possessed ways to respond to the threats directed against them, weaponry provided a crucial way to answer the terror of combat — and, by offering the soldier some response, created another opportunity to cast the battlefield as controllable in some ways, filled with important choices for individual combatants to make. Taking concrete action to answer those threats formed a powerful spur to a soldier's willingness to endure combat. A World War II GI remembered his weaponry as a potent salve to the fear he experienced in battle: in combat, he recalled, he was "scared at first until I started to shoot back."[3] A predecessor in the Civil War recalled that in each battle, "the first shot I fired seemed to take all my fear away and gave me courage enough to calmly load my musket and fire it forty times."[4] That connection between soldiers' attitudes toward their weaponry and their willingness to fight became progressively more sophisticated (and more central to their motivations to fight) as the technology of warfare evolved, precisely because the sense of control it provided became even more critical on the dispersed battlefields of the twentieth century.[5] The possession of a weapon provided those soldiers a physical embodiment of the way their decisions could affect their outcomes. A GI fighting in the European Theater during World War II summarized that connection succinctly. Reflecting on a moment when his patrol surprised a group of Germans, he described the exchange in vivid detail: "The Germans spun around and fired in our direction," and the American platoon returned fire. The GI took careful aim at one German "and slowly squeezed the trigger on my M-1 carbine." The result unfolded "almost like slow motion." Through his sights, he witnessed the German's "head open up as the bullet passed through the left side of his face": a "sickening" spectacle — but, caught up in the heat of battle, the soldier "didn't have a reaction other than feeling I had saved my own life."[6]

In the main, rank and file soldiers of the War of Independence and the Civil War experienced combat as elements of a homogenous unit. Every member of an infantry regiment's companies carried one of a handful of models of smoothbore or rifled muskets. Precious little specialization existed among individual soldiers in a regiment: with few exceptions, each rifleman carried weapons similar to those of his comrades (and, given the relative crudity of the technologies—which allowed comparably few variations in design—often identical to those his enemies carried). Those soldiers utilized basic, exposed-mechanism weapons familiar from their civilian lives, and the simple flintlock and percussion-cap trigger devices of eighteenth- and nineteenth-century weapons facilitated an infantryman's ease with his arms. The common cartridge, a solid lead ball and a pinch of black powder, embodied simplicity. The practical considerations of those muzzle-loading arms helped emphasize the interdependence among the members of a firing line. The lengthy period required to reload the weapons (during which the distracted soldier was painfully vulnerable to enemy fire, without any means to respond) reminded the foot soldier that he was but a member of a larger machine and that the machine's solidarity and integrity provided him with his most effective protections against enemy fire. For the eighteenth- and nineteenth-century soldier, the limitations of the weaponry helped strengthen the motivational effects of literal, shoulder-to-shoulder contact in linear formations.

In the twentieth century's dispersed tactical system, the relationship between soldier and weapon became more complex. A far greater degree of specialization existed among infantrymen on the battlefield: an American World War II platoon always featured a heavier Browning Automatic Rifle, and an infantry company included a wide range of weaponry, from light machine guns to flamethrowers to portable, indirect-fire mortars. That broader range of weaponry differentiated individual soldiers to a much higher degree. Technological advances, particularly the introduction of semiautomatic and fully automatic breech-loading weapons that decreased the lengthy reloading period between shots, along with the increased range and explosive power of twentieth-century ammunition, made the lone GI of the Second World War a far more lethal force than his counterpart in the armies of the eighteenth and nineteenth centuries. But the faster, more complex automatic mechanisms were also far more vulnerable to malfunction; a jammed weapon left a soldier unarmed in the heat of combat, a precariously vulnerable position for an infantryman dispersed from the closed ranks of the firing line. Spread over wider spaces, often out of contact with comrades, troops fighting in the dispersed system had to exercise more individual flexibility and far greater individual initiative than soldiers in the linear tactical system. On the empty battlefield, the soldier's own perceptions of his effectiveness played a critical role in generating the motivation to endure battle; soldiers who believed that their actions in combat could help them survive were far more likely to stay and fight. As the battlefield became a faster and more lethal environment, those perceptions of individual agency assumed greater importance and became more closely linked to the weapons themselves. The individual soldier's relationship with his weaponry grew more nuanced as well. Shoulder arms became much more lethal, but that lethality came at a price: more complex weapons could malfunction in a firefight, and their destructive power some-times targeted the operator himself, rather than the enemy. As the nature of the battlefield evolved to place more and more importance on the individual's ability to maneuver and hide, soldiers often found that the weight and bulk of their arms rendered them slower and more vulnerable targets. Even an act as basic as firing a weapon suddenly involved a new trade-off on the dispersed battlefield, since the noise and flash of the shot threatened to betray a hidden position to a wait-ing enemy. At the same time, the variety of these weapons and their ever-increasing lethal punch provided the individual foot soldier with more decision-making flexibility. The choices available to the World War II GI (what weapon to use, what kind of ammunition to employ, where to aim his weapon, when to fire) increased the individual's sense of potential control over the battlefield environment — and that sense of control in combat (whether real or illusory) provided an impor-tant boost to morale as the demands of dispersion removed many of the tangible physical motivators that had operated to such great effect within the linear system of tactics. Encouraging a soldier to view the battlefield as a controllable environment with which he and his weap-ons interacted provided critical opportunities to channel individual and small-group behavior in militarily useful directions.

Of course, establishing such controllability could work at cross-purposes, as well. On the empty battlefield, a soldier with equipment he believed ineffective or faulty was far more likely to opt out of com-bat situations if the opportunity presented itself. Private Jim Foley, a member of a unit that was overrun by Germans during the Ardennes offensive in the winter of 1944–1945, recalled being trapped in a farm-house with a .45-caliber handgun that, in his words, "couldn't hit a B-29 at ten paces." Rather than attempt to shoot their way out, his small group eventually surrendered to a German infantry unit. Foley

rationalized his group's surrender in part by referring to their inad-
equate armament: though he argued that he "hated to give up like
that," it seemed the best option to the stranded men. The perception that the group's weaponry was light and ineffective ("We didn't
have so much as a BAR in the whole place") led Foley to imagine that
they stood scant chance of success against the more heavily armed
Germans. The imbalance justified their decision to capitulate: "If we
had started shooting, we would have been slaughtered like a bunch of
cattle."⁷ One soldier at Omaha Beach found his weaponry woefully in-
adequate against the enemy fire that combat threw at him. That conclusion further sapped his morale. "My rifle jammed," he remembered
afterward, "so I picked up a carbine and got off a couple of rounds."
But small-arms fire seemed "inconsequential" against the fortified emplacements on the beach, a conclusion that sapped his will to continue
the fight: "There was no way I was going to knock out a German concrete emplacement with a .30-caliber rifle."⁸ The quality of arms (and,
equally important, soldiers' perceptions of the quality of their arms)
affected the combat motivations of twentieth-century infantrymen
dramatically. (Troops' relationship with their weapons affected the
behaviors of even eighteenth- and nineteenth-century soldiers, though
to a lesser degree. Union soldier Lorenzo Vanderhoef complained to
his father about the quality of his company's muskets in an 1861 letter:
"I declare it is too bad that we must use those old things that will kill
more behind them than in front."⁹ Jonas Elliot, an officer in the 102nd
Ohio Volunteers, wrote the following year that the inferior weapons
his men had received impeded their willingness to face battle. Their
guns were infamous Austrian rifles, widely disparaged as unwieldy, inaccurate, and ineffective.¹⁰ Elliott believed the poor reputation of the
weapons and his men's dissatisfaction with them were undermining
their motivation to fight: "The men have no confidence in their guns
& dont want to go into battle with such arms.")¹¹

Regardless of the state of the technology, certain limitations applied
equally to soldiers of each war. Foot soldiers in each conflict had to
carry most of their own arms, ammunition, and rations. That weight
was substantial; as a typical Civil War soldier described his burden,
"my load consisted of my knapsack, with 1 pair pants my fancy shirt,
one pair socks 1 pair drawers 1 pound tobacco 1 Pair boots some paper
an empty haversack, canteen of water, gun, cartridge box, 40 rounds
of cartridges &c &c."¹² Advances in technology (the use of lighter al-
loys, more effective explosives, and innovative artificial fibers) made
the equipment and weaponry of the mid-twentieth century lighter,

more durable, and more effective. The total weight and bulk of equipment a normal-sized soldier could carry into combat varied little from the War of Independence to the Second World War; the destructive power of that gear, however, expanded geometrically between the eighteenth and twentieth centuries. Individually, soldiers of the Second World War represented far more lethal forces than their counterparts in the eighteenth and nineteenth centuries. Technological improvements in portable weaponry afforded the twentieth-century soldier longer range, more accuracy, and vastly higher rates of fire than those enjoyed by foot soldiers in the Continental Army or the Army of the Potomac. New weaponry also altered the types of firepower the soldier could unleash on the battlefield: the explosive power of burst weapons like grenades and mortars lent troops the ability to blanket an area with shrapnel and to hit hidden targets via indirect fire previously available only to full artillery crews in the War of Independence and the Civil War. Portable, shaped-charge antitank rounds gave even the lone infantryman an armor-penetrating capability that soldiers before the twentieth century could scarcely have imagined. Many twentieth-century soldiers internalized the magnitude of the destructive power their weapons afforded. World War II paratrooper Ross Carter described himself and his comrades (bandoliers of ammunition crisscrossing their chests, spare clips in their packs, and grenades bulging in their pockets) as "walking arsenals."[13] Another GI thought that one of his comrades resembled a Mexican bandit, with two bandoliers across his chest, a third filling his cartridge belt, and grenades stuffed into his four jacket pockets. Viewing the portable armory, one comrade pointed out that if the soldier had to go to ground to avoid enemy fire, he would not be able to get back up. The anxious and well-armed infantryman protested, citing his fear of running out of ammunition; his comrades ultimately persuaded him to give up one ammunition belt and the grenades by pointing out that he already carried "enough to kill 240 Krauts, 120 if you miss half of them."[14]

The importance of emphasizing individual agency within the dispersed system surfaced again in the lengths to which the army went to demonstrate to the average soldier the destructive power contained in his weaponry. In one case, instructors marched three soldiers a thousand yards downrange after a firing exercise that tested recruits on their skills with the 60mm mortar. There they witnessed the effects of their barrage. Abandoned junk cars and some temporary shacks had provided a canvas on which to demonstrate the power of the mortar shells. The effects were both obvious and impressive: battered sheds, splintered wood, and a demolished car. The damage testified to the power that could be wrought by a three-man team armed with a 60mm mortar tube. The evidence was, as the soldier recalled more than a half-century later, "an eye-opener." The mortar projectiles did not leave large explosive craters, but the effects of the thousands of pieces of shrapnel they had unleashed appeared tattooed across the junk vehicle: "The front fender of the old car was nowhere to be seen, the hood was punched full of holes, the crankcase busted open, and the rest of the body attached to the chewed-up front displayed ragged holes." The lesson to the soldiers was clear: "We were convinced that nothing and no one could survive within twenty-five or thirty yards of a strike." The demonstration cemented the troops' sense of the kind of destructive power they would carry with them into battle. "That this degree of damage could be caused by one small three-pound projectile from one little smoothbore tube," he recalled, "rendered us sober and silent." That realization, in turn, led to an even more significant conclusion, one that bolstered their collective sense of efficacy: "We felt like we had in our one mortar section the firepower to destroy any and every thing we were likely to encounter."[15] (Tellingly, none of the trainees leapt to the next logical insight: that the enemies they were likely to encounter would bear their own mortar tubes with which to rain down similar devastation on *them*.) No similar up-close demonstrations of their weapons' destructive effects accompanied soldiers' training during the War of Independence or the Civil War. Little time and spare ammunition existed for such luxuries; moreover, reassuring soldiers who would fight from the massed ranks of linear formations was far less necessary. Not only was the individual destructive capacity of those soldiers far more limited, their opportunities to dodge combat once engaged (standing within the "moving box" of the formation and under the ever-watchful eyes of their officers) were comparatively limited. Soldiers spread out and camouflaged in combat frustrated attempts to control their behavior directly. Those soldiers required other reasons to continue their active participation in combat. Tangible demonstrations of the effects their firepower could have in eliminating enemy threats helped cement the connection between going forward and surviving the trial of battle.

Larger loads of ammunition and the more powerful, rapid-firing automatic weapons they fed created that enormous leap in the destructive power of the individual soldier during the Second World War. The standard load of ammunition in the eighteenth-century Continental Army consisted of forty cartridges and three flints; a Civil

War soldier's cartridge box held forty rounds (optimistically termed "forty dead men" by the soldiers who carried it, after the maximum amount of destruction it could inflict); as a rule, troops carried another twenty rounds into combat, stuffed into pockets or haversacks.[16] A World War II rifleman armed with the semiautomatic M1 Garand rifle used ammunition packaged in eight-round clips, each roughly the size of a pack of cigarettes. The substitution of steel-jacketed, smaller-caliber bullets for the Civil War soldier's boxful of half-inch solid lead slugs meant that a World War II GI could carry a dozen or more such clips without significantly more bulk than was contained in the mid-nineteenth century soldier's cartridge box. New automatic and semiautomatic weapons also dramatically increased the number of bullets a soldier could put into the air. Combined with industrial production and modern logistics that provided soldiers with vastly more ammunition than their counterparts in earlier eighteenth- and nineteenth-century armies, the weapons represented a quantum leap in the amount of firepower a single soldier could unleash. A quick-firing Civil War soldier, who might be able to launch three shots per minute in combat conditions, had in his sixty rounds enough ammunition to last fifteen or twenty minutes of continuous firing.[17] Twentieth-century industrial production and delivery freed armies from some of those logistical constraints, allowing soldiers to bring the full might of automatic weapons to bear on the battlefield. A World War II soldier equipped with a semiautomatic M1 could fire off sixty shots, a Civil War soldier's full battle complement of ammunition, in fewer than three minutes. A paratrooper's automatic carbine could fire two thirty-round bursts in less than a minute. Heavier World War II automatic weapons boasted cyclic rates of fire approaching 500 rounds per minute, giving the lone World War II GI greater firepower than an entire hundred-man Civil War company.

Not just the amount but the type of ammunition the GI carried distinguished him from earlier American foot soldiers. Ammunition during the War of Independence embodied simplicity: a smooth, solid lead ball twisted into a scrap of paper with a thimbleful of gunpowder. Late-eighteenth-century bullets were so crude that soldiers could cast their own shot around an evening's campfire using a molding tool small enough to hang from the belt. Some troops used homespun tricks to make their ammunition more deadly, scoring bullets (which then split into four pieces upon impact) or adding jagged bits of iron or rusty nails to their shot.[18] While Civil War soldiers rarely needed to cast their own ammunition after the first months of the war, the bullet itself was only slightly more complicated than the Continentals' solid shot. The Minié ball, that war's iconic projectile, employed a hollow base that flared at the moment of firing to fill the barrel of the weapon and accept the rifled grooves. In the early part of the war, before Confederate armories began the mass production of ammunition, some rebel troops casting bullets found that they could fabricate their own makeshift Minié balls by inserting a stick into the base of the mold as the lead cooled to create a hollow indentation. The Civil War infantryman's cartridge was simple enough that, in a pinch, a soldier could fashion his own in the field using a dash of powder, a ball, and a scrap of paper wrapped together with string.[19] The simplicity of their munitions afforded soldiers of the late eighteenth and early nineteenth centuries an intimate connection with their weapons: uncomplicated and sometimes homemade, their cartridges held little mystery for them.

The mid-twentieth century, in contrast, presented the rank-and-file soldier with a dizzying array of munitions. Soldiers could no longer fashion their own ammunition in the field; modern automatic weapons required mass-produced, factory-made ammunition (brass cartridges filled with steel-jacketed slugs, crammed into spring-loaded clips) to feed the gun quickly and effectively. The wider variety of weaponry required soldiers to master a much larger inventory of cartridges, clips, and ammunition belts. Unlike their counterparts in earlier wars, nearly all of whom used the same interchangeable cartridges, the soldiers of the Second World War had to be able to quickly identify which ammunition was appropriate for which weapon. In addition, soldiers carried different types of ammunition to serve a variety of tasks. Riflemen could choose among regular ammunition, tracer rounds that left a bright streak to aid in aiming, armor-piercing rounds that could penetrate metal plating, and other varieties of ammunition, each designed for a particular task. The vast new array of ammunition available to mid-twentieth century soldiers underscored the options available to the individual in modern industrial war. Some infantrymen became loyal to certain types of ammunition. Harry Arnold, a veteran of the European Theater, recalled that before going into combat, "I made sure that all my clips were loaded with black tip AP (armor piercing) rounds." That was not standard practice in his unit ("Most of the men disagreed with my choice"), but to Arnold the habit had a sound basis in logic: "I intended that no bastard hiding behind a tree, wall, or light vehicle would be impervious" to his bullets. The decision involved a tradeoff; normal ball ammunition did more damage upon impact and had more stopping power, but possessed less penetrating ability than armor-piercing rounds. But it was a tradeoff Arnold was willing to accept: "If AP left a small neat hole through all rather than more serious wounds, I felt assured that, once drilled, most men would not come back for more."[20] The greater choice of ammunition offered more subtle encouragement for soldiers to think about the battlefield as a controllable environment by furnishing yet another decision to make that might reasonably have an effect on their outcome in combat. Other soldiers employed their weapons in decidedly unorthodox ways in order to get them to behave in a particular manner. A GI observed one sergeant who, though he had excelled at target practice while training stateside, had nonetheless determined that the machine gun "was not a precision instrument." Rather, the sergeant believed it was best employed "like a hose, spraying as many locations and troops as possible." To facilitate the spray of the "hose" before combat, he "deliberately held down the trigger to burn up a full belt," an action that violated the training manual, which prescribed brief intermittent bursts to keep both the gun's barrel and its steel jacket cool. The rapid-fire full-belt barrage "left the barrel so hot that cartridges were set off from barrel heat alone, without being struck by the firing pin." In addition, constant use wore away the rifled grooves in the barrel, decreasing the weapon's accuracy such that "at two hundred to three hundred yards, instead of pumping six shots into one enemy soldier, the bullets scattered from the sloppy barrel randomly over several feet," thereby "increasing chances of bringing down more than one and adding to the fear and distress of more enemy soldiers."[21] The new sophistication of twentieth-century arms provided enterprising soldiers with new ways to mold them to their own ends, reinforcing the sense of the battlefield as an environment that responded to individual effort in meaningful ways.

It was not just the ammunition but the weapons themselves that were more complex. Shoulder arms of the late eighteenth century were relatively straightforward affairs: smoothbore muzzle-loading arms that featured only a few moving parts, easy to understand and uncomplicated if exhausting to operate. Most were easy to clean and, when necessary, to repair. Moreover, the weapons featured mechanisms identical in operation to those the soldiers used for hunting or home defense—and in the case of many militia units, of course, the weapons were the very ones the men used in peacetime. The Continental soldier's basic flintlock musket was a simple, if not entirely reliable, shoulder arm: poor workmanship could render as many as half the weapons in a unit worthless, and many muskets refused to function properly in damp or windy weather.[22] Civil War arms, though considerably more reliable, were only slightly more complicated: the mechanism on the nineteenth-century rifles featured an exposed percussion-cap ignition system that was both visible and easy to understand. The infantry weapons of the Second World War, in contrast, were far more complex tools that demanded a correspondingly higher level of mechanical aptitude from its bearer. The soldier issued the trusty Browning Automatic Rifle, one to every squad, also assumed the responsibility for the care and upkeep of its exacting mechanisms; one American marveled at the "tiny springs, wheels, and pins" that constituted the spare parts kit of the high-maintenance weapon.[23] That mechanical complexity produced the automatic weapons' vastly higher rates of fire, but the increased firing speed came at a price: the weapons proved much fussier, requiring constant attention and adjustment. One soldier remembered the particular quirks of the M3 "grease gun." The bolt springs that fed the gun had to be stretched by hand frequently, and though the gun's magazine was designed to hold thirty bullets, "if 30 bullets were loaded the magazine springs would not have enough force to lift the bullets, so I had to load only 20." This fix alone did not guarantee reliable function, either: "Even then I had to empty the magazine every so often and stretch the springs again."[24]

The routine maintenance modern weapons required constituted yet another distraction demanding the soldier's attention even in the heat of a firefight, a diversion that proved fatal in some cases as preoccupied soldiers inadvertently exposed themselves to enemy fire. But the weapons also gave the soldier of the mid-twentieth century an enormous advantage in the number of bullets he could put into the air in a short space of time. Portable indirect-fire mortars provided soldiers even more flexibility in the type of targets they could attack, but they simultaneously demanded that the individual master an even wider range of weaponry and skills. The function of such complex weapons was not immediately obvious to the neophyte, and an untutored soldier with a sophisticated weapon was hard-pressed to get any useful service out of it at all. One GI in the European Theater found himself on Omaha Beach lying next to a mortar and faced with an order from his sergeant to fire it. "I fired two or three rounds and they flew out of the tube," the soldier remembered, "but they didn't explode." The rounds' failure to detonate was easily explained by the more experienced sergeant, who yelled, "Murdoch, you dumb bastard, you're not pulling the firing pins!" His mystification stemmed from the fact that he had never fired the mortar before, but heeding the sergeant's instructions he pulled the pins of the remainder of the rounds. The improvised crew knew "we got some mortar fire on the beach, but with the smoke and my fogged glasses, I couldn't tell what we hit."[25]

GIs of the Second World War could also extend their killing reach much farther than eighteenth- and nineteenth-century soldiers. The Continental musketman firing buck-and-ball could expect some degree of accuracy at about fifty yards, and the projectile would still pack a lethal punch for another hundred yards or so. Soldiers bearing long rifles enjoyed more accuracy: writing after the war, one English officer estimated that an expert rifleman could hit a target at 200 yards and that some might be able to hit targets at 300 yards, though evidence from the conflict itself suggested that those estimates were overly generous.[26] The complicated steps necessary to load the long rifle meant that the rifleman's rate of fire was even slower than his counterpart with a smoothbore. Civil War soldiers, employing large-bore rifled shoulder arms, could deliver accurate projectiles at longer ranges: the model 1861 Springfield rifle, workhorse weapon of the Civil War infantryman, had an effective range between 300 and 400 yards and could still be lethal at distances up to 1,000 yards.[27] World War II weapons, with their vastly higher muzzle velocities, granted the GIs of the mid-twentieth century another leap in both accuracy and range. The twentieth-century soldier's arsenal also included a wider range of weaponry, from grenades and mortars to handheld antitank weapons, which allowed soldiers to assault targets that had been previously impervious to infantrymen. Portable, stable explosives (grenades, bricks of TNT, telescoping Bangalore torpedoes) enabled even a lone infantry soldier to cut barbed wire, attack armored targets, and destroy bridges and buildings, giving a single World War II GI a destructive capacity and flexibility beyond the imagination of predecessors in the eighteenth and nineteenth centuries.

But the World War II GI's increased individual lethality came at a price to the soldiers themselves. The very equipment that made an individual soldier so dangerous to the enemy was also more dangerous to the infantryman who carried it. Some of the weapons were simply more hazardous to operate. Infantrymen of the Second World War quickly learned that some of the equipment in their arsenal was dangerous even when used properly; in this respect, they had more in common with the artillerymen of the eighteenth and nineteenth centuries, who had to exercise extreme caution in the operation of their weapons, than with their predecessors in the infantry. (Firing eighteenth- and nineteenth-century cannon required close attention to a half-dozen precise steps, and any deviation or carelessness put the entire gun crew at enormous risk: powder bags inserted into a muzzle that had not been properly sponged of burning embers might explode, an improperly seated powder charge could produce barrel-splitting pressure when ignited, premature detonation of the cannon might crush an unprepared artilleryman under the wheels' recoil.)[28] For the most part, infantrymen in the War of Independence and the Civil War had little to fear from operating their shoulder arms; even defective weapons, which threatened to explode the barrel when fired, usually did so harmlessly. A private in the 7th Minnesota Volunteers who wrote the manufacturer of his rifle in 1863 after his barrel burst noted that his was the second in his company that had proved faulty; neither mishap, apparently, had injured the owner.[29]

World War II infantrymen quickly discovered that some of their weaponry did indeed pose a real threat to the firer. Donald Greener, a World War II armored infantryman, remembered that when triggering the M1 antitank weapon (that war's famous bazooka) "one had to wear an eye and face shield and gloves" to provide protection from the flames that flew out of both ends when the weapon fired.[30] Even soldiers who did not typically operate the weapon had numerous chances to witness its deadly power, and the way that power was sometimes directed against the bearer. Soldiers who spent much time near the front witnessed a variety of scenes that supported pessimistic conclusions about the inherent dangers of modern weaponry: a half-century after the episode, an American GI recalled vividly a decapitated German soldier whose body still cradled a *Panzerfaust*, the German antitank grenade launcher. The American inferred from the scene that the man's own weapon "apparently had misfired and blown up in his face."[31] The World War II soldier's arsenal included more explosive weaponry as well. The flamethrower, which mounted tanks of jellied gasoline on the wearer's back, was particularly vulnerable: tracer rounds that hit the reservoir could easily explode the volatile contraption. One soldier at Omaha Beach remembered just such a moment in grisly detail: "I heard a blast and saw that a man wearing a flamethrower had been hit, and the fuel tank was on fire." It was a hideous scene ("The man with the flamethrower was screaming in agony") and as the victim dove overboard to extinguish the flames, his observer noted that even the soles of his boots were on fire.[32] Grenades, a standard part of the combat infantryman's arsenal, posed a similar danger as soldiers tumbled across the battlefield with small explosive charges clipped to their gear. Phil Hannon, a soldier at the Second World War's Battle of the Bulge, described one close call: "A grenade I had hooked to my jacket fell off. When it hit the ground, the pin fell out." Only the weight of the grenade pressed on the handle kept it from detonating.[33] Again, it was not the objective danger of the grenade itself (surely such mishaps caused an insignificant number of casualties during the Second World War) but the soldier's perception of that danger that mattered. Rumor of such an accident, the type of story likely to spread rapidly among anxious and superstitious GIs, might reasonably cause a soldier to view the weapons he carried with suspicion. The danger a combatant's own ammunition might pose to his safety was another relatively new threat to the modern infantryman. While in certain isolated cases the eighteenth- or nineteenth-century soldier's ammunition might endanger his life (some wounded soldiers at the 1864 battle at the Wilderness, for example, had the gunpowder in their cartridge boxes ignited by the brush fires that spread across the wooded battlefield and were shot to death by their own ammunition), such cases were exceedingly rare.[34] The soldier of the Second World War had to worry about not just the threat posed by enemy bullets but also the threat presented by his own ammunition.

The physical bulk of the World War II GI's weaponry and ammunition constituted another kind of new danger. While less cumbersome than the equipment carried by his predecessors in the eighteenth and nineteenth centuries, its size and heft restricted some of the soldier's movements, rendering him a clumsier, slower target. Moving quickly on the battlefield was of relatively less importance to foot soldiers of the War of Independence and the Civil War; locked into linear formations, few had an opportunity to dodge or avoid enemy bullets. Remaining steadfast under fire, rather than rapid movement across the battlefield, was the critical task for those soldiers. For World War II GIs employing move-and-fire techniques to limit their exposure to the storm of enemy fire, however, a speedy dash across the beaten zone was critical, and a few seconds might mean the difference between survival and death. Cumbersome gear thus became an obstacle, rather than an ally, in the struggle to remain alive. One overburdened American soldier complained that "the ammunition belt I had was too large and couldn't be further shortened; the weight of it pulled down over my hips" and made maneuvering difficult—a handicap that threatened the life of a soldier in the dispersed system to a much greater degree than a soldier fighting in the linear system.[35] Limits to mobility created a tradeoff between individual lethality and the soldier's own survival. And the bulk of modern weaponry could endanger soldiers' lives in other ways. Private First Class Harry Parley carried an eighty-pound flamethrower on his back at Omaha and nearly drowned under the weight of the device: "I stepped off the ramp into a deep pocket in the sand and went under completely." With no footing and the weight of the flamethrower on his back, Parley found it nearly impossible to surface. Aware that he was going to drown, Parley made what felt to him like a "futile" attempt to unbuckle the flamethrower's harness and freed himself just before it killed him.[36] (Paratroopers in particular learned to be especially wary of jumping over areas that contained waterways. The massive amount of equipment they carried—everything the self-contained soldier might need to sustain himself for several days until reinforced and resupplied, including main and reserve chutes, rifle, ammunition, combat knife, gas mask, rations, map, and first aid kit—quickly became waterlogged and could drag a freshly landed airborne soldier under quickly and permanently: killed not by enemy fire but drowned by his own gear.)

For the World War II GI, even the infantry soldier's most basic act of war, firing his weapon, could endanger his life. The flash and noise produced by firing a shot could reveal a hidden soldier's position, drawing enemy fire. Eighteenth- and nineteenth-century weapons produced noise and smoke, of course, but the exposed linear formations and volleys of the battlefields in the War of Independence and the Civil War placed little premium on stealth or concealment. For soldiers standing in dressed lines aligned on prominent banners, extra smoke and noise did not render them any more vulnerable. For soldiers on the twentieth century's empty battlefield, where staying hidden often constituted a critical part of staying alive, betraying a camouflaged position invited potentially deadly retaliation as the telltale flash and noise that a discharged weapon produced signaled a soldier's position to enemies in the area. (Armored units struggled with this problem even late in the war; a tanker in northwest Europe grew frustrated that every time the crew fired the gun, they "revealed our position with both smoke and flame" since the flash "could be seen for miles.")[37] The antitank bazooka had a similar effect. As one user remembered, "Once you've fired a round and given your position away, their infantry are on you like a swarm of bees." The sense of resignation such a realization engendered was profound: "You might as well put your head between your legs and kiss your butt goodbye."[38] Tracer rounds, which left bright streaks to help a gunner see where his shots fell, helped the firer home in on the target, but at a cost: enemy observers could follow the tracers back to their source to ascertain the location of the user.[39] Even ostensibly benign equipment sometimes carried a hidden threat. Troops going ashore in North Africa in 1942 learned to discard their sunglasses, a seemingly innocuous item, because the lenses reflected the sun's glare and could alert enemy gunners to the presence of appealing targets.[40]

Revealing the infantryman's position was only one way a weapon could endanger a soldier's life. Jammed or broken weapons left the soldier vulnerable and presented him with an unpleasant choice: remain helpless with a nonfunctional weapon, unable to return enemy attacks or provide suppressing fire, or take attention away from combat long enough to fix the weapon. Misfires have plagued soldiers since the advent of gunpowder, but the modern automatic weapons of the twentieth century were far more sensitive and much more prone to mishaps, since they obtained the high rate of fire through complex mechanisms that were far more susceptible to dirt, moisture, and temperature. A World War II GI allowed that "all of the tactical skills in the world went for naught in machine gunning if the weapon didn't work," and keeping the delicate mechanisms functional "meant fastidious cleanliness." The soldier produced an extensive catalog of ways in which the conditions of battle could dirty the gun and render it useless: "As the hot barrel cooled, moisture and resulting rust were drawn into every nook and cranny. Burnt primers and powder fouled the action, stopping it or leaving it gummy and sluggish. The front barrel bearing and booster, a cup-shaped device at the muzzle, loaded up with carbon and had to be removed frequently and painstakingly scraped clean."[41] Some weapons earned a particular reputation for jamming no matter how well-maintained by the operator, and soldiers (of course) found those shortcomings frustrating and obnoxious. One infantryman remembered that his weapon, the pressed-metal M3 "grease gun," jammed exceedingly frequently, as even "the tiniest piece of sand would jam the cylinders."[42] Even the trusted Browning Automatic Rifle (BAR) was susceptible to problems. Pete Opengari, who fought in the European Theater with the 36th Infantry Division, recalled having "a BAR that wouldn't fire" due to "sand or dirt in the chamber." The fix was novel: Opengari handed the gun to a comrade, who "took a leak right into the chamber." While unusual, the remedy proved effective, as Opengari remembered that the gun "did its duty that night."[43] Other weapons misfired in different ways; one soldier in another division recalled his unit's flamethrower malfunctioning in the heat of a firefight, spewing not fearful flames but a harmless milky fluid that, in his understated words, "failed to impress the defenders."[44]

Even a properly functioning weapon produced some disadvantageous, and often downright dangerous, effects for its user. Deafness was the most common complaint: the effect of shooting off a dozen rounds in quick succession with the weapon in firing position close to the cheek left the soldier with ringing in at least one ear. The temporary deafness of the battlefield has been a problem for soldiers of every conflict since the advent of gunpowder. A soldier at the 1863 battle of Stones River lamented the "incessant din of musketry" in battle and "the ringing in one's ears" that it produced.[45] But for the twentieth-century soldier, scattered into dispersed formations and far more dependent on sound cues to provide him with information about the enemy's location and actions that might help save his life, the danger posed by such temporary deafness loomed much larger.

Over time, the growing sophistication of the infantryman's basic tools rendered the relationship between soldier and weapon more complex. In the armies of the War of Independence and the Civil War, the musket clearly represented a tool meant to serve the soldier. That relationship was not as clear by the middle of the Second World War; the individual and the tool had become a more symbiotic pair, inviting the soldier to wonder whether his weapons and gear were there to serve him or whether he was there to serve the weapon.[46] That confusion manifested itself in the ambivalence many World War II soldiers felt toward their equipment. One of Bill Mauldin's cartoons captured the absurdity soldiers felt about this reversal perfectly; in the cartoon, an artilleryman weathering a downpour under a flimsy piece of newspaper while his own helmet and raincoat protected his artillery piece explained to a curious passerby, "I'd ruther cover th' gun. I won't have to dry *myself* with a oily rag."[47] The soldier of the mid-twentieth century might rightly regard his weapon with suspicion: the same tool that made him dangerous as a warrior also made him a slower, more appealing target, laden with explosive ammunition that might render an opponent's lucky shot disastrous.[48] One result of the increasingly symbiotic relationship between soldier and equipment was the way that weapons often became personified. A trusted, battle-proven weapon might win soldiers' praise for its reliability or adopt, in some circumstances, talismanic powers. One World War II paratrooper remembered his squad's .30-caliber machine gun, which remained with the unit from training in England all the way to Bastogne, outlasting several gunners. In spite of its bullet holes and battle scars, "the men thought of it as almost a living member of the squad."[49]

The larger variety of weaponry in the twentieth century had other effects on the soldier's sense of the battlefield. Soldiers of the Civil War, for example, usually carried shoulder arms nearly identical to those employed by their opponents. (Indeed, the Confederacy's poor capacity to fabricate its own arms early in the war, and the resulting determination its quartermasters brought to stripping battlefields of arms and ammunition, meant that Confederate soldiers often carried the exact same weapons as their Federal opponents.) There was so little difference between the standard arms of the Continentals and those of the redcoats they opposed that few American soldiers thought to make comparisons between their arms. Soldiers of the mid-twentieth century, by contrast, frequently compared their equipment with the enemy's in search of cues about their own degree of preparedness and effectiveness. On occasions, soldiers conducted these comparisons directly: scavenging afforded enterprising troops access to discarded enemy weaponry, and curious soldiers occasionally put recovered enemy weapons in head-to-head trials against the corresponding American equipment. Roscoe Blunt remembered performing such a trial in a ruined church using shattered statues as targets: "The idea was to see whether the Colt .45 or the 9mm German P-38 Luger was the most accurate pistol." After exhausting numerous clips, the men agreed that the German weapons were superior "by a wide margin."[50] Especially savvy soldiers understood that superiority could be a relative measure: one World War II foot soldier remembered that the M1 was not as accurate as the German Mauser at 500 yards and that even at 200 yards the Mauser had a slight edge. "But," he recalled, "we usually fought in the range of twenty to one hundred yards." At those distances, the M1 was as accurate as the German weapon and, he believed, "much faster." Soldiers who could not get their hands on enemy equipment used other sources as their basis for comparison. Sound cues provided the most ready basis for judgment: a German machine gun "opened up and its rate of fire was so great that it sounded like a piece of cloth tearing." The noise of the German weapon rang in sharp contrast to the corresponding American weapon, which soldiers described as "more of a chatter." Infantryman Thomas Rosell judged the American machine guns similarly wanting: "I can remember the difference in machine guns, the very fast, high . . . cyclic rate of the German MG-34 or MG-42." The U.S. gun, meanwhile, made a less-impressive, staccato "tat-tat-tat-tat."[51] Another GI made similar comparisons, remembering that "you could always count the shots from our guns as separate shots, but Jerries' MGs and machine pistols were just a liquid rattle, with one shot so close on the heels of the other that it sounded like cloth ripping or 'brrupbrrupp.'"[52] Another soldier judged his M3 "grease gun" similarly deficient: though he had expected it to shoot fast, it instead launched its bullets slowly and "sounded like an old man trying to cough." The gun's sight was so wide (an indication of its poor accuracy) that its bearer imagined that he could see an entire battalion through it.[53]

A soldier whose army-issued weapon was observably deficient to his enemy's suffered a corresponding drop in morale. The sense that his own odds of survival were comparatively lessened through no shortcoming of his own combined with a reduced sense of responsibility to the military system that did not take pains to send him into combat with well-designed, well-built equipment. Such sound cues (or even the lack of them) often created a sense of fear or awe regarding the enemy's capabilities. American soldiers believed that shells from the German 88, surely the most-feared enemy weapon of the war, traveled so fast that they arrived ahead of their sound: soldiers bracketed by 88s, conventional wisdom held, never even heard the explosion that killed them. Such a fearsome reputation was dangerous because it gave lie to troops' sense of the battlefield as a controllable environment: after all, if a soldier received not even a split-second warning of an impending projectile, what action could he possibly take to save himself? Comparisons with enemy equipment could also have a detrimental effect on troops' confidence in the military system as a whole. Soldiers in combat naturally wanted to believe that they were the best-equipped and best-trained and that they therefore had the best chance of survival. Evidence to the contrary suggested, however subtly, that the chances for staying alive remained frustratingly, and frighteningly, slimmer than necessary. That kind of speculation was not limited to equipment at hand but extended to larger issues, from the overall system of supply to the sense of technological superiority. Bill Davidson recalled hearing the telltale *whoosh* of a German Me-262 jet fighter at the beginning of the Ardennes campaign, and the sound (he never saw the plane itself) had a depressing effect on his morale: "Our side didn't have any jets in Europe. I hadn't heard the 'whoosh'—nothing but the steady propeller roar of our P-47's and P-38's—since the first few days after I had had the misfortune to land on Omaha Beach in Normandy." The noise of the German jet started Davidson toward a frustrating conclusion about the overall state of the Allied war effort:

"To myself I said, 'What the fuck is going on? Where the hell are *our* guys?'"[54]

In many cases, mere possession of a weapon provided a soldier a kind of talismanic effect, lending an aura of power. Again, it was often the perception rather than objective reality that lent this feeling of potency; in many cases, the power of such perceptions could override a soldier's own logical reasoning. As he and a comrade approached a wooded position being shelled, one American recalled that "Taylor stopped and took out his pistol. I took out mine too." Intellectually, the soldier had "doubts about the effectiveness of a .45 against an 88-millimeter cannon," but he concluded that the action was reassuring nonetheless, as "it was nice having something in your hand."[55] World War II veteran and author Paul Fussell recalled deriving a similar feeling of power from his weapon in an early firefight: leaping for a stand of M1s, he luckily grabbed one with a round in the chamber. Pushing off the safety, Fussell emptied the eight-round clip. That small triumph gave Fussell a lasting confidence in the M1: "From that moment on, I carried an M-1. I had great faith in that weapon."[56] Imbuing weaponry with special power was hardly unique to soldiers of the twentieth century. In 1864, one Civil War soldier sent his father money for the purchase of a new rifle. Money was unimportant to the soldier ("I sent for the best quality and I expect it will be very high priced but the price I care nothing about for I want the best rifle that can be got"). But he did include one special request: a man the family knew in town was to select the weapon personally: "He is the best judge of a rifle there is up there. And, he will get a good one."[57] In the Second World War, some soldiers' rifles served both as a symbol of their military efficacy and set them apart in other ways. As one of his platoon's six scouts, Leon Standifer remembered receiving a Winchester rifle, which he preferred to the Springfield rifle issued to the platoon's other members. Standifer appreciated the beauty of the weapon itself (the stock, in particular, was of handsome figured walnut) but appreciated its meaning even more: "The Winchester became my symbol for being a scout." Standifer took special care of the weapon, passing it on to a comrade after sustaining a wound.[58]

But the talismanic effect that possession of a weapon provided could cut both ways. Paratroopers relieving a position at Bastogne during the Battle of the Bulge came across a group of grimy, dazed soldiers retreating from the advancing German wave. One member of the 101st approached a withdrawing soldier and asked him to join the relieving force; the man declined. The paratrooper then asked if the man would relinquish his rifle, arguing, "You don't need it and there are a lot of us who don't have weapons." The retreating soldier's response to the troopers' entreaties illustrated the power the simple possession of a weapon could have over an individual's psyche: the young man "clutched his M1 close to his body, his arms wrapped tightly around it as though he were cradling a baby, and started running down the road." Even after putting distance between the paratroopers, "he kept glancing back at us in wide-eyed terror, fearful that we might pursue him and take his rifle."[59]

In order for an army to fight effectively, it demanded soldiers who were willing both to endure the risks inherent in bearing arms and to discharge their weapons in a variety of traumatic and unpredictable circumstances. For the army to remain effective, the soldiers' view of his weapons needed to remain epiphenomenal rather than dependent upon the specifics of a given situation. The most useful way to accomplish this task was to condition a soldier to channel perceptions in combat through a matrix of ideas that encouraged a sense of his own individual agency. Old Sarge exemplified this type of thinking. Advising incoming soldiers to clean their rifles "once a week on general principles, whether or not it has been fired," Sarge likened such regular maintenance to paying periodic life-insurance premiums. "You don't know when your heirs will collect the insurance," Sarge counseled, "but you do know that you have to pay the premiums on time to protect them. There's this difference: when the time comes to collect on the proper care of your rifle *you* are the one who benefits."[60] Sarge's homespun analogy exaggerated the soldier's own role in preserving his life. The explicit instruction—*care for your weapon regularly to ensure it works properly*—echoed a more subtle, implicit message the soldier might recall from training: *there are things we will teach you to do that will help you save your life.* Too, there was Sarge's canny assertion that a properly functioning weapon served the soldier primarily, and the goals of the military system only secondarily. The twentieth-century GI's decision to pull the trigger at the critical moment was the product of the army's long campaign to condition his behavior and decisions. But at that critical moment, the conditioning exerted only an influence: actually pulling the trigger remained an individual act, just as it had been for American soldiers in the eighteenth and nineteenth centuries.

Pulling the trigger in combat was crucial. But for many military philosophers, the way soldiers interacted with their weapons remained of secondary importance to the way soldiers interacted with friendly troops. That was the heart of Marshall's findings: in *Men against Fire*, he argued that the soldier is "sustained by his fellows primarily and by his weapons secondarily." Given a choice, Marshall held, soldiers of the Second World War "would rather be unarmed and with comrades around him than altogether alone, though possessing the most per-fect of quick-firing weapons."[61] In battle, a unit's combat effectiveness depended in large part upon trust: the assurance that comrades were all doing their part to protect the lives of the men in the unit and to achieve an objective. The complexities of those relationships with comrades, their evolutions over time, and their connection with indi-viduals' willingness to fight, form the subject of the final chapter.

6

Comradeship

Joshua Lawrence Chamberlain compiled a remarkable record as a soldier during his Civil War service in the Union Army. Over four years, the professor who left his post at Bowdoin College to join the Union Army fought in half a dozen major engagements, anchored the vulnerable left flank on Little Round Top during the Battle of Gettysburg, survived serious wounds at Fredericksburg and Petersburg, and received the Confederate surrender in 1865 as a brevet major general. After the war, he reflected upon the factors that drove the men under his command in battle after battle. Several dynamics, from honor to masculinity, were at work, but at bottom, Chamberlain wrote, it was "love, or bond of comradeship—'Here is Bill; I will go or stay as he does.'"[1] Similar opinions surfaced repeatedly in other veterans' ruminations on the factors that propelled them in combat on very different battlefields. Joseph Dougherty, fighting in Europe during the Second World War, argued that while training and discipline drove a soldier to the battlefield, "camaraderie makes him perform in fire fights."[2] S. L. A. Marshall drew on scores of interviews with American troops in the Second World War and the Korean conflict to create his theories about the relationship between comrades and combat performance. The "thing which enables the infantry soldier to keep going," he concluded in his pathbreaking work *Men against Fire*, "is the presence or presumed presence of a comrade."[3]

Testimonials to the power of the bonds of comradeship to motivate soldiers in combat have appeared over and over in veterans' letters, diaries, and memoirs. And they recur commonly in American soldiers' reflections regardless of the particular conflict in which the authors fought. References to the intense, familial connection between members of a combat unit are nearly as common in the eighteenth century as the twentieth. Though the language is slightly different, the

172

Civil War scholar need not look too far to find a soldier of the mid-nineteenth century whose ideas about the strength of the bonds that formed between members of a combat unit anticipate the Vietnam veteran cited in Jonathan Shay's *Achilles in Vietnam:* "It's a closeness you never had before. It's closer than your mother and father," closer than relationships with brothers and sisters.[4]

In the middle of the twentieth century, scholars studying the factors that led soldiers to endure the battlefields of the Second World War focused on the bonds of affection and mutual interdependence between soldiers who suffered and fought together as the primary source of combat motivation. Three influential studies that appeared at the end of that conflict (Marshall's own *Men against Fire*, Edmund Shils and Morris Janowitz's "Cohesion and Disintegration in the Wehrmacht in World War II," and Samuel Stouffer's *The American Soldier*) distilled those notions into a single concept, primary group cohesion.[5] In the latter half of the twentieth century, primary group cohesion became the orthodox explanation for soldiers' often seemingly irrational behaviors in combat, employed by scholars and researchers to understand actions in conflicts as diverse as the British troops in the Pals Battalions of World War I, American and German units clashing in the European Theater of Operations during the Second World War, the American Civil War, the Falklands War of 1982, the Gulf War, and the twenty-first-century conflicts in Afghanistan and Iraq.

Two factors in particular lend credence to the group cohesion thesis. First, the notion that soldiers fight primarily for one another echoes the firsthand observations of countless generations of soldiers. Reflecting on the factors that motivated them in life-and-death battlefield situations, large numbers of veterans concluded that they fought mainly for the men alongside them. Second, the notion is intuitively satisfying. The natural appeal of the idea that soldiers fight principally for one another is perhaps most apparent in its enduring grip on the popular imagination, a hold that has endured for centuries. Act IV of William Shakespeare's *Henry V* finds the warrior king encouraging his outnumbered troops on the eve of battle by addressing them as a "band of brothers," bound together by their shared experiences and sacrifices in combat. Four hundred years later, Steven Spielberg's 1998 film *Saving Private Ryan* featured the title character (a young World War II paratrooper whose three brothers all have been killed in combat, and who is to be withdrawn from battle as a result) refusing to leave his fellow soldiers at their post. Instead, he instructs the officer leading his rescue party to tell his grieving mother that "when you found me, I was here, and I was with the only brothers I have left. And that there's no way I was going to leave them."⁶ Stephen Ambrose's immensely popular books on the combat soldiers of the Second World War (one with the revealing title *Band of Brothers*, which was adapted into an award-winning television miniseries) at once reflected and fed the popular idea that the bonds of comradeship enable soldiers to overcome their fears and fight. As the introduction to this book noted, however, several serious problems exist with employing the primary group thesis as a blanket explanation for behavior in battle. Though consistent with soldiers' own anecdotal observations and intuitively appealing, a large body of historical evidence resists the argument that cohesion is the principal source of combat motivation. That evidence undermines the primary group cohesion thesis in several important ways.

Historian Omer Bartov's 1991 *Hitler's Army* offered one of the most direct challenges to the notion that primary group cohesion generates combat motivation. In his examination of the German army on the Second World War's Eastern Front, Bartov noted that the group cohesion thesis was difficult to sustain in battle due to primary groups' "unfortunate tendency to disintegrate when they are most needed." The heavy toll of combat, he observed, killed off precisely the comrades thought to support the group's combat motivation. Mining Shils and Janowitz's own data, Bartov argued that those authors' findings reversed the chronology of events: German units on the Eastern Front actually fought with more determination in the final three years of the war, though by that point the savage fighting in the East had inflicted casualties in German combat units ranging from one to two times the unit's original strength.⁷ That level of loss would destroy the cohesiveness of the units, as old soldiers perished and the bonds linking the members of the unit began to dissolve. If the orthodox explanation were correct, these Wehrmacht units should have become dramatically less willing, if not altogether unwilling, to fight as their camaraderie eroded precipitously. That was clearly not the case in Bartov's sample: late in the war, German units fought with even more tenacity after enduring massive casualties. As Bartov put it, the presence or absence of strong primary groups alone could not explain troops' combat motivation.⁸

Examples from American armies echoed Bartov's analysis. Robert Rush's examination of the 22d Infantry Regiment during the savage fighting in the Hürtgen Forest in 1944 noted that the unit lost more than nine of every ten men during the nineteen days it spent in combat. The "band of brothers" explanation of combat motivation simply could not account for those soldiers' dedication during the brutal weeks of combat.[9] Joseph Glatthaar's *General Lee's Army* included a similar case from the Civil War, relating the story of Virginia private William A. Burke, who forged powerful bonds with the soldiers in his unit (particularly within his mess, the small group of a dozen or half-dozen soldiers who shared cooking duties and tented together) that undoubtedly helped him face the horrors of close-quarter Civil War combat. But after the disastrous Confederate assault on the third day of Gettysburg (in which his unit, the 18th Virginia, lost more than three of every four soldiers engaged), Burke learned that the last surviving member of his mess, one of his brothers, had perished in the attack. Yet even the total liquidation of his primary group, Glatthaar notes, did not destroy Private Burke's willingness to fight: he continued to serve with the remnants of his regiment and returned to battle after Gettysburg.[10] Combat journalist Ernie Pyle reflected on the transience of primary groups during the Second World War, noting that "it is natural to be loyal to your friends, and I feel a loyalty to the First Division, for I lived with it off and on for six months." It was, however, "a sad thing to become loyal to the men of a division in wartime. It is sad because the men come and go, and other new ones come until at last only the number of the division is left." In the final analysis, the First Division appeared to Pyle as but "a numbered mechanism through which men pass": the division would continue to exist, though many of his friends in it would not.[11]

A second weakness in the group cohesion thesis first emerged from experimental psychologists and sociologists. That research demonstrated that tight primary group cohesion could in fact work against organizational goals, emboldening group members to resist authority imposed from above—precisely the kind of opposition that is devastating in hierarchical organizations like military units. Philip Zimbardo's famous Stanford Prison Experiment (in which the experimenter randomly designated volunteers either "guards" or "prisoners," allowing the guards wide latitude in writing and enforcing rules for their charges) is perhaps the best-known example of this phenomenon. Over the ten days his mock detention center was in operation, Zimbardo found that the subjects playing the prisoners were fairly compliant at the outset but began to challenge the authority of the subjects cast as guards as the inmates came to know each other better, to appreciate their shared plight, and to form cohesive bonds with one another.[12] Historians studying the real-life incidence of soldiers deliberately "fragging" their own officers during the Vietnam War discovered that these episodes were often collective efforts—group members pooling a bounty on the head of a targeted officer, for example—in which the cohesiveness of the unit lent a powerful group sanction to the criminal action.[13] Military sociologist Charles Moskos estimated that 80 percent of Vietnam fraggings were the result of group action. "It is an irony of sorts," Moskos reflected, "that the primary group processes which appeared to sustain combat soldiers are close cousins to the social processes which underlay the vast bulk of fraggings in Vietnam."[14]

A final problem with the group cohesion theory is nearly as intuitive as the appeal of the theory itself. The group cohesion argument derives much of its power from the strength of the emotional bonds it taps: soldiers are willing to risk their own lives and fight, the argument holds, because of the deep bonds of comradeship they feel with their fellow soldiers. But these powerful emotional bonds can have other, counterproductive consequences for military units. Strong emotional ties among soldiers do not necessarily, or even logically, lead to a desire to fight. An infantryman who charges into battle primarily to support a comrade with whom he shares a strong bond of affection may find his attention shifted away from the unit's military objective if, for example, that friend suddenly suffers a combat wound. Roger Little's work on buddy pairs during the Korean War confirmed that many soldiers considered their loyalty to comrades a higher priority than their loyalty to the organization or to the mission. In interviews, soldiers revealed that, if wounded, they expected their buddies to stay and render aid, although orders and the demands of the mission dictated that uninjured soldiers press the attack rather than stay behind to comfort the injured.[15] Employing the group cohesion thesis as the main explanation for combat performance overlooks the ways in which the very strength of the primary group bonds can lead to individual actions that are at cross-purposes with military goals.[16] Together these three weaknesses help explain why scholars have yet to establish a reliable correlation between group cohesion and group performance beyond the anecdotal evidence supplied by soldiers themselves.[17] The previous chapters in this book suggest another problem with the orthodox explanation: given the vast differences in the experience of combat for the average foot soldier from the eighteenth century to the twentieth, it is fundamentally ahistorical to expect that a single phenomenon can explain behavior in such widely disparate landscapes. At the very least, the mechanisms of the bonds of trust and affection must have operated differently depending on whether the members of a unit were packed tightly into lines shoulder-to-shoulder with their comrades or dispersed and hidden from one another's sight. Marshall himself, whose work helped establish the foundation for the group cohesion thesis, acknowledged as much in *Men against Fire* when he wrote that in twentieth century battle "the prevailing tactical conditions increased the problem of unit coherence in combat."[18]

The accumulated weaknesses in the group cohesion theory have led some social scientists to look more closely at the phenomenon of cohesion as it applies to combat soldiers. The search for a compelling link between cohesion and the willingness to endure enemy fire has suggested a distinction between social cohesion, the degree to which members of a group identify with one another and enjoy spending time together, and task cohesion, the degree to which members of a group share a commitment to some collective goal. Groups can be socially cohesive without exhibiting a high degree of task cohesion, as any number of "clubby" but ineffective organizations demonstrate; similarly, groups can possess a low level of social cohesion (with some members actively disliking others) but still demonstrate a high level of success in achieving collective goals—as indicated by scores of championship sports teams, among other examples.[19] In the context of fighting units, it seems, it is task cohesion that is the critical factor encouraging soldiers to face fire; social cohesion, while not without its uses, is neither necessary (as demonstrated in the work of Bartov, Rush, Glatthaar, and others) nor sufficient (as argued in the work of Zimbardo, Little, and others) to get soldiers to endure the trauma of battle. The notion of task cohesion also provides a useful new way to consider the power of the bonds among members of a unit in combat. From the perspective of the military, a group's task is its mission: capture those cannon, assault that ridge, hold a particular crossroads until reinforced. But from the perspective of the individual, the overriding task in battle was often far simpler and more urgent: *stay alive.* Achieving that goal forced soldiers to work together, albeit in very different ways as technology transformed the linear battlefields of the eighteenth and nineteenth centuries into the open, empty battlefields of the mid-twentieth century.

Many veterans and scholars acknowledged the importance of simply remaining alive as a powerful component of the bonds linking the members of a combat unit. Studying the behavior of American troops in Vietnam, sociologist Charles Moskos dismissed phenomena noted by earlier scholars as the "semi-mystical bond of comradeship," arguing instead that deeply pragmatic needs and desires (particularly the instinct for self-preservation) had underlain these bonds, rather than allegiance to some idealized "band of brothers."[20] The Vietnam veteran quoted in Jonathan Shay's work testifying to the more-than-familial bond between the members of his platoon acknowledged that truth in his interview, noting that soldiers could not attend to even the most basic hygienic needs without the support and protection of a comrade: "Y'know, you'd take a shit, and he'd be right there covering you." The veteran's conclusion highlighted not emotional bonds but the most basic natural instincts for self-preservation: "We needed each other to survive."[21]

The dynamics of those relationships changed over time, but some constants persisted. Infantry soldiers seldom faced combat alone; nearly all went into battle as part of a larger unit. No matter how well trained, well led, or well armed, a lone infantry soldier rarely represented much of a force on the gunpowder battlefield. Soldiers wielding muzzleloaders on the battlefields of the eighteenth and nineteenth centuries simply could not discharge their arms quickly enough to assemble an intimidating fire; their projectiles were accurate only at limited ranges, and while reloading soldiers stood excruciatingly vulnerable to enemy fire. The infantry soldier of the mid-twentieth century was equipped with semi- or fully automatic weapons and thus was capable of generating a volume of fire orders of magnitude greater than his predecessors; nonetheless, he found himself much more vulnerable to the increased quantities of enemy fire leveled against him amid the "storm of steel." On the empty battlefield, individual soldiers were most effective (and often only effective) when working in concert, combining suppressive fire with movement in coordinated teams. Out of necessity, the American armies of the War of Independence, the Civil War, and the Second World War channeled the efforts of training and of combat leadership toward creating groups of individuals who could work together to maximize the military power brought to bear against the enemy in combat.

The blunt realities of the battlefield brought the American soldier's relationship with his comrades into sharp focus. In different ways, the tactics of both the linear and dispersed battlefields reinforced the importance of comrades. But the precise nature of the mutual support soldiers provided their comrades changed dramatically as the decades passed. Infantrymen in the War of Independence and the Civil War required the close presence of their comrades to form their line of battle. As the basic unit of both attack and defense, the battle line provided physical reassurance to men trembling in fear (the "touch of the elbow" mentioned so frequently in the letters and diaries of Civil War soldiers); helped compensate for the shortcomings of eighteenth- and nineteenth-century weapons (coordinating fire by volley); and made flight under fire extremely difficult, since the men were packed into tight lines from which a route of escape was both physically challenging and virtually guaranteed to be observed by neighbors in the ranks, as well as by watchful officers and noncommissioned officers authorized to employ physical coercion to keep the soldiers in formation.

The dispersed tactics of the mid-twentieth century's empty battlefields altered those dynamics dramatically. Gone were the dressed lines of the eighteenth and nineteenth centuries; close-packed formations created too rich a target for opposing gunners' automatic fire. As enemy salvos forced infantrymen to seek cover, individuals could no longer reassure fellow soldiers with their physical presence nor check up on their behavior directly during battle. Instead, interactions with friendly soldiers helped fuse the unit into a fighting force in markedly different ways. Comrades acted as each other's eyes and ears, scanning the field for potential dangers and communicating this information to their fellows. They also provided the covering fire that made movement possible for friendly soldiers by forcing enemy infantrymen to keep their own heads down. The literal touch of the elbow was absent, but other factors (the presence of crewed weapons like mortars, for instance, that required teams of men working together to operate) reminded soldiers that they remained fundamentally dependent on one another for both their military effectiveness and, more important to the individual participant, their survival. In addition, comradeship provided soldiers of all three wars with a critical emotional encouragement. The extremes of combat magnified the natural need for psychological support, friendship, and validation. For the soldier at the front, exposed constantly to the rigors of battle and isolated from family and home, the world often extended only as far as the immediate unit.[22]

At bottom, this brand of comradeship among soldiers in combat revolved around mutual respect, interdependence, and commiseration based on shared deprivation. It coalesced into a powerful kind of trust that helped soldiers endure the trauma of combat. But this critical trust among friendly soldiers assumed different forms as the specific dangers of the battlefield and the requirements of different tactical systems placed new demands on those relationships. Within the linear tactical system, trust fundamentally revolved around the promise that fellow soldiers would stay in the firing line despite the acute dangers of enemy shot and shell and the powerful natural instincts urging them to flee. On the dispersed battlefield, infantrymen's trust depended upon the knowledge that fellow soldiers would perform their jobs, and perform them competently, even when individual soldiers could not monitor the actions of their comrades directly. During each of the three wars, the military system provided a variety of formal and informal controls to keep soldiers at their tasks. Awareness that survival depended in large part upon the presence and performance of comrades ultimately bolstered the formal and informal controls and helped steel soldiers to stay and fight at the moment of contact.

On the battlefields of the eighteenth and nineteenth centuries, one of the soldier's strongest incentives to stay and weather enemy volleys was the controlling interdependence among individuals serving together on the firing line. The brutally straightforward dynamics of linear tactics created compelling reasons for individual members of a company or regiment to remain physically cohesive. On the linear battlefield, the sense of communal safety the line offered was far from illusory, and the touch of neighbors furnished physical proof that the line had not broken. The psychological support provided by nearby friendly troops often proved nearly as useful. Union foot soldier Jonathan Stowe described the effect of cheers from friendly units as "electric." "It thrills the senses," he wrote in a letter describing the sensations of battle, "to feel assured by enthusiastic shouts that plenty of friends are present to back you." That those cheers supplied such a boost spoke to a more altruistic desire to support fellows in the line. But a second part of the comment illustrated the powerful disincentives to flee that the linear system provided simultaneously. Spurred by the cheers of their comrades, Stowe reported, soldiers on the Civil War battlefield went "forward with a *rush* lest *your pride taunt you of cowardice.*"[23] Not altruism but pragmatism operated in this regard: within the closely packed ranks of the linear battlefield, cowardice could be observed and punished (immediately in some cases, later in others). In that way, the close quarters of the linear battlefield provided troops engaged in a firefight with powerful discouragement to leaving. Those disincentives ranged from the explicit coercion of file closers ordered to shoot fleeing soldiers to the implicit threats of ostracism that met soldiers found shirking their duties in combat.

Given the way that American units in the War of Independence and the Civil War were recruited by state and often from a single town or county, those indirect threats against social status could carry nearly as much weight in battle as more direct threats. Soldiers in many eighteenth- and nineteenth-century units found themselves

serving alongside men they knew from home. In the opening months
of both the War of Independence and the Civil War, it was common
for large groups of men from a town to enlist en masse, and throughout both conflicts geography helped define infantry regiments. The
longstanding bonds of home proved a potent tie between individual
soldiers: assured that stories of their cowardice would follow them
back to their civilian lives, armies assumed that eighteenth- and
nineteenth-century soldiers would be less likely to break and run in
front of men they knew. Stephen Crane's protagonist in his 1895 novel
The Red Badge of Courage felt this pressure acutely, as he imagined his
whole regiment asking, "Where's Henry Fleming? He run, didn't 'e?
Oh, my!"[24] Another Union soldier wrote his father to report on the
behavior of a neighbor from home, who had "always shirked his duty
off on the other boys" and had not stood guard or picket detail since
entering the service. The man's avoidance of duty had won him an
ignominious nickname in the unit, "the 'Flunky' of Company K." Pre-
dictably, the company flunky performed no better in battle than he
did in camp. "As far as bravery is concerned," his colleague wrote,
"there is not but very few in the Co. but what has more courage going into battle than he has." (The author was surprisingly indulgent
in his letter, instructing his father, "You need not let anyone know
much about this letter," but other soldiers in both armies were not
as lenient.)[25] Infantrymen of the eighteenth- and nineteenth-century
battlefield expected tales of dishonorable behavior to cast shadows into
postwar life. In addition, the tightly packed ranks of the close-order
battlefield made it physically difficult for soldiers to escape the firing line once battle began. For a soldier in the middle of a formation,
there was simply little place to go but forward: friendly soldiers at
either elbow and a line of troops behind him made retreat a dubious
proposition. Crane's Henry Fleming discovered this truth of the linear formation in one of the novel's battle scenes, when "he instantly
saw that it would be impossible for him to escape from the regiment.
It inclosed him." Surrounded by other Union soldiers marching in
lockstep, Crane's hero found himself trapped inside what was essentially "a moving box."[26]

Besides the powerful motivational dimension of these tactics, close-
order formations proved critical to preserving a unit's physical integrity and thereby to maintaining its military effectiveness. Massing fire
was one of the most important functions the line fulfilled. Packing
the men into lines also helped compensate for the lengthy firing times
of the muzzle-loading weapons of the War of Independence and the

Civil War. The rank-and-file foot soldier spent the best part of his time under fire loading and preparing to fire his weapon. The linear system helped provide some protection to the group by keeping up a steady stream of lead that promised to help disrupt enemy fire somewhat. While the first volley, and possibly the second, erupted simultaneously, the pattern of fire became more ragged from then on as the fastest shots in the company loosed more rounds than their slower-firing counterparts. That flow of bullets kept pressure on the opposing line and helped guarantee that enemy soldiers were equally harried in their own attempts to load, aim, and return fire.

Breaking under fire thus represented enormous danger not only to a unit's integrity and its ability to achieve military goals but to the safety of the individual soldiers who constituted it. At once, flight erased the individual soldier's sense of physical reassurance, threatened to spread panic in the ranks, and eroded the fire support critical to the individual infantryman's task. As with the patriot soldier who made the "instantaneous" decision to flee the Battle of Camden in 1780 once it seemed that everyone else "was about to do the same," the appearance of panic in close-ordered ranks could prove immediate, infectious, and disastrous. Poor communication aggravated the danger, as soldiers engaged in a fight usually had no way to know whether departing friendly soldiers were redeploying to offer more effective support or simply fleeing altogether, leaving the position exposed as a result. The individual soldier's resolve within a linear formation thus resembled a variation of the political scientist's familiar prisoner's dilemma: every individual in the unit was safest if the entire unit maintained the integrity of the line and kept up its fire. But if the line appeared on the verge of rout, it was better to be the first soldier to flee (when the relative chances of escape were better) than the last, who stood precariously vulnerable to an oncoming enemy. Deciding how to respond in the chaos of battle, the soldier's instinct was often to err on the side of his own safety. A soldier who fled the Camden battlefield recalled hearing the noise of cannon fire from the field shortly after his departure, but confessed he did not know its orientation at the moment he retreated. As it turned out, the sounds emanated from friendly cannon, and the soldier later surmised, "Had we known, we might have returned."[27]

Because breaking under fire presented such an immediate and dire threat to the safety of the entire unit, punishments within the linear system had to be swift and severe. Fleeing from battle constituted the most serious offense, of course. But even smaller violations merited immediate redress, lest indiscipline become a habit that ultimately created panic among members of a unit in actual battle. Some minor disciplinary infractions warranted a pay stoppage; more serious violations in conduct demanded corporal punishment. One soldier charged with disorderly conduct during the Civil War found himself at three months' hard labor with a ball and chain affixed to one leg. Acknowledging the seriousness of the punishment, one of his fellows nonetheless appreciated its necessity: "It is rather hard, but I am glad of it."²⁸ The nature of combat in the linear system, and the necessity of maintaining the integrity of the line, led soldiers inclined toward steadfastness to welcome a draconian system of penalty for indiscipline. Infractions, veterans of the firing line understood, endangered every member of the unit if they threatened the solidarity of the formation.

Disorder in the linear system was one thing; cowardice was quite another. Widespread cowardice in actual battle never became a staple of American soldiers on eighteenth- and nineteenth-century battlefields (in general, soldiers whose fears overcame them before or during combat skulked around the rear of their regiments until the danger passed), but the more serious danger that cowardice in battle presented the rest of the unit led to massive antipathy or outright hatred toward those who shirked in combat.²⁹ As one New York volunteer at Gettysburg put it, "I had a good notion to fire at cowards. I would about as soon shot one as a reb."³⁰ Because cowardice in combat put the entire unit at risk, some deserters received capital sentences. Other regiments utilized different punishments: the 2nd Vermont infantry branded two deserters, "one on the right hip and the other on the left shoulder with the letter 'D,'" in November 1863, and another two the following week.³¹

A soldier breaking in combat on the linear battlefield endangered lives beyond his own immediate neighbors in the line. Josiah Atkins witnessed a neighboring brigade of Pennsylvanians break at the 1781 Battle of Green Spring; the Pennsylvanians, once routed, could not be formed again, "by which," he observed, "they lost their *field-piece* & we the ground."³² The Pennsylvanians' refusal to stand fast imperiled both Atkins's own brigade (by destroying the integrity of the American line and opening gaps for the enemy to exploit) and the rest of the patriots on the field (since the captured artillery piece could be turned against the Americans at Green Spring or pressed into British service later). The sense of cohesion in battle among friendly soldiers within the linear system of tactics thus constituted not just a vague, brotherly affection but also a very specific kind of trust: trust that the soldier's neighbors in the line would not crack when the enemy's fire became blisteringly hot and every natural instinct urged them to flee. The pronounced unit pride that emerged in many of the veteran units of the War of Independence and the Civil War embodied one manifestation of this trust. Unit pride expressed more than simple vanity; it carried the reassurance that the unit had proved itself able to stand up under the rigors of combat, even (or especially) when other units had proved incapable of bearing the pressure. A strong sense of unit pride and elitism was therefore more than the hard-earned satisfaction in the performance of one's unit, though that was certainly a component; in addition, pride in a particular unit helped reinforce soldiers' sense that their relative chances of survival were increased by dint of the unit's superior combat ability. (That is one reason armies went to such great lengths to stimulate unit pride and to remind soldiers of their unit's combat history. The flags and streamers, the cheers, the lore: all tied the individual to comrade, neighbor, state, and nation. In addition, they provided the soldier a sense that others had passed the test of combat, and that he, like them, could measure up.) A Union soldier writing home in the autumn of 1863 revealed just such a powerful sense of pride in the accomplishments of his division. Noting that the Confederates had taken up a strong position across the river in his unit's front, he nonetheless indicated that his army was "bound" to drive the rebels back on Richmond and boasted that "we are the boys that can do the thing up right."[33]

Another manifestation of that pride appeared in the tensions between eastern and western armies among the Union forces of the Civil War. The two armies' records (eastern armies suffered an almost unbroken string of setbacks in the war's first two years, while western armies enjoyed a nearly uninterrupted series of successes) led to catcalls and taunts whenever western troops came into contact with their eastern counterparts. Soldiers in these meetings could not help but notice a difference in the cultures of the two armies. Rice Bull suggested that eastern generals made more of a show of military pomp with sashes, swords, and ribbons displayed prominently on "brilliantly dressed staffs." Among western armies, Bull "never saw such a gaudy show; there they seemed to avoid show of any kind."[34] The decorations did little to impress western veterans: within the broad fraternity of soldiers, the quickest way to earn the respect of other friendly units was to demonstrate courage and competence on the battlefield. That was certainly Henry Welch's experience fighting with Hooker's Corps; by 1864, that unit had been transferred to the western theater and had undone much of its earlier ignominy, and Welch proudly noted that "the western men dont try to run us down so much as they did." His corps' performance in battle had changed some of the westerners' minds, and "they begin to think that the Potomac men can fight after all."[35]

Sensitivity to the abilities of comrades in the regiment, and its effects on constituent members, surfaced again in the Union Army two and a half years into the Civil War. As 1863 drew on, the looming expiration in spring 1864 of the three-year enlistments that original volunteers had signed threatened to free large blocks of the Union's best infantrymen to return home. Desperate to maintain its veteran troops, the government authorized a reenlistment program to encourage those soldiers to sign on for additional service rather than return home when their initial obligation expired. Soldiers who agreed to extend their terms received a variety of incentives, including bounty payments and a furlough home. One of the incentives was communal: if three-quarters of a regiment agreed to reenlist, the entire unit received special recognition as a "Veteran Volunteer" unit and a guarantee that its members would continue to fight together as a regiment under their old colors. Veterans who reenlisted but could not convince the requisite three-fourths of their comrades to do likewise were folded into other existing units. Many veterans found the prospect of amalgamation distressing; one wrote to his father of his concern that he might leave his hard-fighting unit for another regiment, and that "perhaps that Regt has disgraced itself in some previous engagement."[36] Read one way, such sentiments reflect a deep sense of nineteenth-century honor and a desire to avoid association with troops who had incurred some dishonor in battle. But it is equally possible to read such statements in part as a pragmatic response to the dynamics of the linear battlefield and a realistic desire for self-preservation. After all, units primarily disgraced themselves in combat by breaking under fire, and units that collapsed in battle left their individual members vulnerable when they ran. The reward of remaining with comrades under the old colors thus accompanied the muted threat of being placed in a unit of inferior troops, with dire implications for the individual soldier's survival.

The dispersed tactics of the Second World War introduced profound differences in the ways that friendly troops interacted with one another in battle. Yards of empty space separated friendly soldiers, and the men employed camouflage, improvised holes, and cover to conceal themselves from enemy gunners' sight. The natural herd instinct that proved so valuable in the linear system, driving men together and keeping them together when bullets flew, became a liability in the dispersed system, since groups of men became appealing targets for enemy guns. Instead of brightly colored uniforms and banners clearly marking the locations of friendly troops, GIs wore drab uniforms deliberately designed to reduce their visibility. Together, those techniques proved remarkably effective in disguising troops: in training, one observer was surprised at how well the uniforms hid friendly soldiers, even when he knew their locations. As his companions sheltered behind trees and bushes, he discovered that "even at such short distance it was hard to pick them out because of the blending of the uniforms with the ground."[37] In some cases, soldiers could not hear their comrades, either. Sometimes, the distances between them were simply too large; further, GIs quickly learned to keep quiet lest their noise give away a hidden position and invite retaliation. Unlike infantrymen of the eighteenth and nineteenth centuries, who used shouted commands, drumbeats, and bugle calls to convey information on the battlefield, the dangers of combat frequently forced soldiers of the Second World War to rely upon silent hand signals to communicate.

Soldiers of the dispersed system remained dependent upon their fellows for support in battle, just as their predecessors in the linear system. But the precise nature of the support they provided one another on the battlefield evolved into a more complicated brand of interdependence. Standard move-and-fire tactics aimed to neutralize the defender's automatic-fire advantage by rotating assignments among members of a squad. While one element applied suppressive fire to force enemy gunners to keep their heads down, another element took advantage of the brief lapse in defensive fire to move to a new sheltered position. Then the elements switched roles: the moving element laid down suppressing fire while the firing element took to the run. Thus interactions among friendly troops became much more dependent upon the specific competencies of their comrades: a squadmate's aim and timing, for example, assumed a critical new significance for the nervous GI about to break against an enemy emplacement.

In many cases, dispersal also bred confusion. That confusion heightened the danger from what became known in the twentieth century as "friendly fire," since the cover, concealment, and greater distances of the empty battlefield made it far more difficult for soldiers to distinguish friend from foe. (Fratricidal incidents were by no means confined to the battlefields of the twentieth century: one Yankee soldier remembered the fate of a friendly regiment of cavalry that pushed forward too eagerly to harass the enemy's rear and found itself in the path of its own army's cannon and received "the benefit of our fire and thereby lost four horses." Had the cavalry not retired, he judged, "they would have been destroyed by our own artillery."[38] No men were lost in the exchange, but other units in other battles were not as lucky.[39]) While isolated losses to friendly fire incidents occurred on eighteenth- and nineteenth-century battlefields, the conspicuous uniforms and flags designed to reduce confusion in battle, and the closer quarters in which opposing armies engaged, made those incidents comparatively less frequent. On the dispersed battlefields of the twentieth century, such losses became tragically common.

In addition, the enormous size of the armies in the Second World War meant that the twentieth-century GI was part of a much larger organization whose magnitude dwarfed the insignificant individual self. Both world wars witnessed enormous growth in the importance of artillery, and by 1944 the army's dependence on both long-range guns and air power to deliver ordnance in support of ground troops added a new, extended group of friendly soldiers upon whom the frontline infantryman had to rely. Significantly, these artillerymen and airmen were increasingly removed further and further from the infantry-man's world at the front: GIs' lives depended upon the abilities of men who were miles away, and whom the foot soldiers never met and often never even saw, to an unprecedented degree. The support of air and artillery was indispensable on the modern battlefield—and GIs paid an enormously high price when these distant units performed poorly.

Though the size of the frontline soldier's support system grew considerably in the Second World War, many individual infantrymen felt more isolated within the dispersed system compared with soldiers who were part of the smaller armies and battlefields of the eighteenth and nineteenth centuries. While the overall size of the armies involved had swollen enormously, World War II combat appeared to be waged by "a small group of men in a corner of the view, almost isolated from the whole," in the words of one World War II soldier. Unlike the enor-mous mobs clashing in the open in the eighteenth and nineteenth cen-turies, members of a World War II platoon were "infrequently able to see their comrades further along, or even the enemy across the way." The physical isolation was disturbing, but the sense of psychological isolation could be even more damaging: "The small group perceives itself alone, is concerned with the few square yards surrounding it."[40] And that isolation bred doubt as to the effectiveness, if not the very presence, of comrades' support. The realities of the dispersed system, which by design hid soldiers from one another, aggravated those feelings of isolation.

Not all of the changes in the nature of infantry combat worked to undermine the individual soldier's motivation. Spread out and often hidden from view, panic could not spread as quickly among the members of a dispersed platoon as it did in the close quarters of the linear battlefield. But the relative isolation of the dispersed battlefield created a host of new problems. Chief among them was the rifleman's lack of information about his comrades' location and status. Spread into foxholes and hidden from view, the soldier's fear of being left alone and vulnerable worked to erode his morale. Surrounded by silent, camouflaged comrades, soldiers were often uncertain whether their units remained in position or had moved out, leaving the individual behind and vulnerable.

Not everything about the infantry soldier's interactions with friendly troops changed over these centuries. Tight bonds continued to form between comrades-in-arms, just as they had on the battlefields of the War of Independence and the Civil War. Like their eighteenth- and nineteenth-century predecessors, twentieth-century GIs found that the close quarters of army life, the sheer amount of time they spent together, and the unifying effects of shared risks accelerated the formation of bonds of comradeship among members of small combat groups. In that respect, the testimony of World War II GIs closely echoed observations of earlier soldiers. A private interviewed in the spring of 1944 characterized the men of his squad as his "special friends" and his sergeant as his closest companion, a relationship cemented by the fact that the two "bunked together, slept together, fought together, told each other where our money was pinned in our shirts." The deprivations of wartime created a strong sense of communal property; as the private put it, "Whatever belongs to me belongs to the whole outfit."[41] Other GIs had similar experiences in their units. Henry Giles spoke fondly of his outfit, where each member would "do anything in the world for you, and for each other," from sharing airmail stamps to donating the last pair of dry socks. Packages from home were understood to be community property. "Usually the guy who gets the package has only one piece of candy, like everybody else, or one cookie or piece of cake." Even cigarettes, scarce and always in demand at the front, passed among the men: "Nobody goes off and lights up by himself. He lights up, puffs it down a piece, then gives the butt to somebody." It was, Giles observed, "a grand, grand outfit."[42] Other American units displayed similar selflessness. When one battalion surgeon appeared with supplies, he brought only a single bottle of whiskey for the entire company; the men, disappointed at the meager stock, nonetheless realized that "it was the best he could do," and appreciated his gesture of goodwill. What happened next was a testament either to genuine selflessness or to the informal sanctions that had developed within the group. Although every man "thirsted for the whole bottle," one witness remembered, each scarcely touched it to his lips "for fear the next man wouldn't get a taste." A few soldiers ("some of the drinkingest men in the army") abstained entirely to ensure their comrades got a share.[43] Here, too, it is possible to read more than simple brotherly affection and altruism into the behavior. A strong element of pragmatism pervaded those actions: these were the men, after all, upon whom the soldier's life depended when under fire. Signaling an unwillingness to share might also imply a lack of concern for others that could resurface with disastrous consequences in the midst of a firefight. As a veteran interviewed by Stouffer's team during the Second World War recalled, "One of the things a fellow learns in the Army is to make friends with anybody, and hold that friendship."[44]

In battle, the teamwork necessary to carry out the tasks of the dispersed tactical system replaced the literal cohesion of the linear battlefield's shoulder-to-shoulder formations. The nature of World War II infantry combat reinforced dependency on the other members of a squad and emphasized the importance of well-practiced communication. House-to-house fighting, for example, required a well-developed ability to anticipate the actions of comrades: the key, one soldier wrote, was "to know what the rest of your squad is going to do," to react to comrades so efficiently that the enemy could not regain his balance.[45] In other cases, the practicalities of the battlefield provided concrete reminders of the ways in which combat bound the men of a unit together. Paratroopers readying for a jump learned this lesson as they strapped on their gear; some of the equipment could not be donned alone and required the help of other troopers to fasten properly. World War II GIs who suffered through the bitter winter of 1944–1945 learned the same lesson more painfully: the woolen gloves the soldiers wore became wet and dried out repeatedly, causing deep splits to appear at the joints of the soldiers' fingers in the process. The cuts made it difficult to perform fine tasks like working pants buttons or lacing up leggings. As one soldier noted, "Frequently men whose hands weren't affected had to perform those tasks for others whose hands were rendered powerless."[46] In such ways, their gear reminded soldiers of the need to pull together to accomplish a task, whether the task was the mission itself or the individual soldier's simple survival. Pragmatic and altruistic bonds between comrades fulfilled other critical military functions on the dispersed battlefield. Where armies had once depended upon officers to monitor troops' combat behavior directly and administer threats and punishments personally, the dispersed system rendered that practice difficult, if not impossible.[47] Part of a soldier's motivation was the genuine desire to help his friends; another part was concern at letting them down, at not appearing man enough to fight.[48] Because the army's direct control over soldiers' behavior waned dramatically on the empty battlefield, the power of disgrace aimed at soldiers who refused to go in grew even more important in the twentieth century. Within the dispersed system, the bonds of comradeship served as an informal brand of discipline that helped bolster the army's weakening control over its infantrymen.

Communal enforcement, often merely implied within the linear system of tactics, became by necessity more overt on the dispersed battlefield. In *How to Get Along in the Army,* Old Sarge instructed new conscripts not to depend on their NCOs to enforce discipline: "Don't get the idea," he lectured, "that it's 'hard-boiled noncoms' who are going to boot your aching bones out of your bunk and drive you to your duty." Instead, a sense of masculine honor would replace the sergeant's ever-present eye: "What's going to drive you, aches or no aches," Sarge continued, "is the knowledge that if you don't hold up your end you'll be marked as a sissy and a quitter, a softy and a goldbrick, by the other men of your own outfit."[49] Other bits of Sarge's advice emphasized the individual restraint imposed by concern for one's standing among the members of the unit. In many cases that restraint focused on the reader's manhood: as far as the aching muscles of boot camp, Sarge warned, "complain about them and you'll be pegged as a cry-baby." Sarge offered similarly straightforward advice for dealing with "no-goods" or "bums" within the outfit. Usually, he suggested, "a unit can make life so unpleasant for him that he'll either transfer or go 'over the hill.'" It was an imperfect solution ("No company commander likes to have desertions on his organization's record") but wartime sometimes required pragmatic measures. It was better, in Old Sarge's estimation, "to toss out a bad apple than to have a whole barrel spoiled." Sarge's lesson encompassed several layers: beneath the practical advice for dealing with slackers lay an implied threat of ostracism for the reader who might be contemplating the role of "bum" or "no-good" for himself. In extreme cases, unit solidarity would serve as a more extreme, if unofficial, form of discipline. In the case of a man who refused to shower, for example, Sarge promised that "his outraged comrades will drag him to the showers and forcibly administer the necessary scrubbing." (As if to emphasize the two-pronged effect of such punishment, Sarge noted with understatement that the "procedure is both humiliating and unpleasant.")[50]

Concern for one's image and reputation proved a powerful motivator for World War II infantrymen. Worried at earning a reputation as a "slacker" if he fell out during a long march, one exhausted GI new to combat forced himself to continue as he "clumsily snapped up my gear and tottered forward again on feet that sent searing pain throughout my wracked body."[51] The power of shame proved effective even on outcasts. Harry Martin, Jr., a private in the 106th division (and the outfit's self-described "sad sack,") received a chance to escape infantry duty altogether by an examining physician, who offered to excuse him if only Martin could come up with another disability to go with his bad vision. As he recalled, the doctor was "definitely on my side and I was moments away from getting out of the infantry, which I wanted so much." But at the critical instant, Martin's nerve failed, and he "just could not give him the answer he was looking for." His behavior resulted, Martin believed, from a combination of shame, conscientiousness, and self-preservation: "I think I felt guilty about leaving the men I had trained with for the last four months. The men of the 3rd Squad were the best there was. I guess I really did not want to leave them."[52]

The bonds of affection, obligation, and interdependence within American World War II units served as a powerful motivator and source of social support. But the relationships among soldiers could also work at cross-purposes with the unit's military goals. Frequently, the demands of the mission required men to act in ways that ran counter to those bonds of friendship. In particular, soldiers often had to stifle their natural instincts toward friends once the bullets began flying. Most soldiers understood, at least on an intellectual level, why this was so. One soldier who was knocked unconscious by a German grenade recognized why his buddies left him behind and continued with the attack: "It was the right thing to do," he noted, reasoning that "you should never stop an attack to look out for the wounded or the dead—if you do, you most likely will become one of them." The fast pace of the modern battlefield dictated that "you've got to keep pressing forward or you'll lose."[53]

Trust among comrades was no less important a part of comradeship on the empty battlefield than it had been on earlier linear battlefields. But by the middle of the twentieth century its specific incarnation had evolved in response to changing tactics and technologies. On the dispersed battlefield, trust revolved more directly around a comrade's competence: the knowledge that a fellow would perform his task capably even without being monitored directly. The individual soldier's own survival depended upon that crucial support from his fellows. The absence of reliable and competent comrades represented a significant obstacle to the formation of cohesive bonds among soldiers. Because their own lives depended so keenly and so obviously on the actions and judgments of their fellows, individual foot soldiers demanded capable comrades with the demonstrated ability to act swiftly, decisively, and effectively. That was the heart of one soldier's objection to fighting alongside a "loner" from his outfit who had not trained with the unit. The soldier expressed concern that the loner, a man named Lehrer, could not be trusted in a pinch, arguing that "he won't carry his part of the load when things get rough." It was not necessarily concern regarding Lehrer's nerve ("He has the guts to do a job if it benefits him,") but the fact that he was not a "team man." Another soldier in the outfit agreed, noting that the squad was like a family, and that only members of the family could be trusted to risk their lives for each other in the heat of battle. Confessing that the breach in confidence ran both ways (none of the soldiers was certain they would risk their own lives for Lehrer, either), they nonetheless noted that Lehrer turned out to be "useless" in combat.[54]

When successful, the sense of trust in the group's collective ability created a useful sense of efficacy that boosted morale among battle-tested units. But the nature of trust based on familiarity created problems as well: combined with the high turnover of the Second World War, it threatened the cohesive bonds that connected the soldiers of a combat unit. The army's system for handling replacements, dribbling new soldiers individually into depleted units rather than replacing battle-ravaged outfits entirely, represented a persistent annoyance to frontline soldiers. One GI of the 99th Infantry Division cited the army's personnel procedures as adequate, for the most part, but singled out the "policy of emptying whole divisions stateside to fill up ranks of decimated overseas battalions" as particularly demoralizing.[55] New men fed into experienced combat units usually had difficulty becoming integrated into the outfit. Obviously, the new men lacked the common experiences that bonded the veterans together; equally as important (but less obvious), veterans had little faith in the green soldiers' combat abilities and resented the fact that the inexperience of these unproven soldiers endangered veterans' own lives in battle.

Part of the reason that new troops suffered so badly in combat, particularly in the industrial warfare of the twentieth century, stemmed from the steep learning curve in battle. The fast pace and great complexity of mechanized combat placed enormous demands upon infantry soldiers, and while they were, on the whole, far better trained than their predecessors in the wars of the eighteenth and nineteenth centuries, the risks to which they were exposed were also far more varied, and the margin for error much slimmer. Only real battle experience could teach the soldier the scores of tricks necessary to stay alive: where enemy fire was likely to come from, how to dig an effective foxhole quickly, when to take action and when to stay put. The necessity (not to mention the difficulty) of learning these lessons in actual combat explained why new men "usually got hit first, in bunches," as one soldier recalled. Because of the replacements' increased tendency to die or become wounded soon after entering combat, experienced veterans had less time and inclination to get to know new soldiers. There was an emotional component to this reluctance, along with a practical dimension: since "most of them died or were wounded badly enough to be sent home during our first hours in combat," forming bonds with green soldiers posed a higher-than-average risk that a veteran would suffer the trauma of losing yet another friend to enemy fire. On the other hand, if the new men could survive their first three days in combat, "it seemed they stood a pretty good chance of lasting much longer — sometimes long enough to eventually become one of the old men."[56] Only those who survived the initial baptism by fire were likely to form cohesive bonds with their platoon-mates, a result of both their shared suffering and the veterans' appreciation for their new comrades' skills. Resentments toward new soldiers also created rifts within units. When new regulations in late 1944 lowered the requirements for new airborne soldiers, veterans in those divisions took exception to the so-called one-day troopers, men who enlisted in the paratroops long after the units' legendary exploits in Normandy and Holland and, presumably, after the most dangerous missions had passed. A veteran trooper referred to these new replacements as "invaders of our select order," and experienced soldiers begrudged the new troopers their jump pay and wings — privileges the veterans agreed the new men had not earned. To original members of the unit, the presence of the one-day troopers served as a sign that their elite ranks had been diluted: the airborne soldiers who once had proudly been set apart from the masses of infantry "were reduced to taking in one-day paratroopers to flesh out our depleted ranks."[57]

The importance of comrades' abilities in keeping the World War II soldier alive and achieving a unit's goals explained much of the GI's careful attention to the bonds that formed among soldiers. Faith in the proven combat ability of one's fellows led to a heightened sense of elitism among experienced frontline units. A strong sense of unit pride developed among many infantry units of the Second World War, just as it had among the armies of the War of Independence and the Civil War. In many cases, the effect was the same: a feeling that membership in one's own unit conferred, by dint of a particular group's collective expertise and ability, correspondingly better odds of survival for the individual. A soldier who fought in the European Theater of Operations (ETO) made this connection unambiguous, arguing that a unit's chances in combat were both controllable and dependent upon ability: "In combat, the company would survive only if its riflemen were capable, confident, and well supplied."[58] Frank Bacon, a Marine who fought in the Pacific Theater, applied the same logic when he elected to volunteer for the Marines: "I knew that they didn't take just anybody—they were pretty selective. And I was interested in associating myself with people who I felt were worthy of getting to know well."[59] Harry Arnold took a similar kind of pride in his company: as he wrote, the men's "sense of well being was accompanied by a sense of pride—we had survived where lesser men would have crumbled."[60]

The enormous size of the American armies of the Second World War allowed a soldier to imagine his membership in the brotherhood of combat as a series of ever-larger concentric circles: squad, platoon, company, division, and so on, all the way through membership in the infantry itself. That realization brought tears to the eyes of one rifleman in the ETO as he marched past a group of support troops, one of whom shouted, "Go get 'em, Infantry!" As tears rolled down his cheeks, he experienced a feeling that recalled the electricity Jonathan Stowe felt in Civil War combat when adjacent units cheered in support: "no fear and tremendous pride."[61] Indeed, the uniting bond of the immediate unit could supercede the overarching influence of the military system in general. Loyalty to the unit, immediate and tangible, frequently overshadowed loyalty to the army itself, distant and abstract. Just before their jump in support of the Normandy invasion, men of the 101st Airborne Division received a special patch identifying them as members of the First Allied Airborne Army; the patch was to be worn on the right shoulder, above the American flag. Many troopers in Donald Burgett's unit "refused to wear the new patch because of strong loyalty they felt for the 101st Airborne Division and its 'Screaming Eagle' patch." A few sewed on the patch but placed it below the American flag.[62]

Scores of infantrymen in the War of Independence and the Civil War commiserated and bonded in part because of the close quarters, shared trauma, and deprivation of soldiering. That universal empathy was not as simple in the Second World War: the sheer magnitude of the conflict, with its enormously larger armies, multiple branches, and vastly higher degree of specialization, meant that many soldiers believed the burdens of fighting at the front were distributed inequitably. That sentiment bred enormous resentment from infantrymen, who remained all but unanimous in the view that the foot soldier's lot was, without exception, the worst. Such bitterness cropped up repeatedly: it was sometimes humorous (as in the Bill Mauldin cartoon in which a tanker with a smoking barrel reassures GIs cowering in their holes, "We'll go away an' stop botherin' you boys now. Jerry's got our range.") but usually far more grave.[63] Frontline soldiers reserved indignation for nearly every branch that did not, in their eyes, share fully in the misery of life on the front (airmen whose status earned them cozy bunks, warm meals, and hot showers when they returned from sorties, or tankers with their "moving foxholes") and even greater ire for ground soldiers in rear areas who were spared the brutality of combat. One GI remembered traveling through France with truckloads of "dirty, grimy, bearded heroes of the front." In the French town of Verviers he and his comrades encountered freshly scrubbed and newly shaved garrison soldiers, who regarded the filthy frontline men with stony expressions. The reception, "sans compassion or cheer," of these garrison soldiers cut the frontline fighters deeply, soon compounded by a further indignity when the returning soldiers learned that they "could not use the shower facilities," despite the fact that the battle-weary troops had not showered for three months. Years later the injustice still stung, and the GI wished for a way to make the garrison troops "fully cognizant that they existed for one reason, and one reason only—to support the line troops one hundred percent."[64] The brutal and unglamorous lot of the frontline foot soldier, and the conviction that only other veteran troops truly understood the particular plight of the combat GI, also bred a powerful us-against-them attitude. Acknowledging that riflemen ranked "at the absolute bottom for prestige," a GI nonetheless pointed out that foot soldiers possessed a "Queen of Battle attitude." "The rifleman," he argued, "was simultaneously the dregs of the company and the glorious knight who would protect his leaders against all threats." The rest of the massive Allied military machine was essentially the infantry's support: "Everything else in the army, navy, and Air Corps existed to get the infantry up there to win the war."[65]

Particularly hard-fought combat units frequently grew bitter at the sense that they bore a disproportionately large share of the fighting burden even among frontline units. Called up in December of 1944 to help contain the German salient at the Battle of the Bulge, members of the American 101st Division felt that they had already borne too large a share of the load, and they resented what they perceived as the necessity of bailing out the infantry yet again. Pointing out that the 101st had already spearheaded the invasion of Europe and the liberation of Holland, one trooper gave voice to the unit's dissatisfaction: "Jeezus Kee-rist. . . . Why can't the infantry take care of themselves?" The airborne's disgust rippled steadily back from the front line: from anger at the line infantry for letting the Germans push them back to anger at the rear-echelon troops in 12th Army Group sitting on their "fat asses," "drinking wine and playing soldier with the local ladies while ordering us to do their fighting." Members of the 101st "felt that a whole lot of people weren't pulling their weight and that we were carrying the load for them."[66]

Only in exceptionally rare cases did frontline riflemen allow that it was sometimes safer for friendly troops at the front than in rear areas. During the Italian campaign, for instance, an American paratrooper noted that the allied forces at Anzio found themselves "surrounded on three sides, with the sea at our back." For the first time during the campaign, "there were no safe back areas," and even the rear-echelon soldiers who usually gloated over their relatively luxurious existence found themselves under fire from the heaviest batteries the Germans possessed. That meant that the front line ("where a man never felt safe") was about as safe as the rear beach areas. Frontline soldiers sent rear to shower "were always in a sweat to get back," since "we felt safer up there." At the front, soldiers had to deal only with small arms, mortar fire, and artillery shells from 75-, 88-, and 105-millimeter guns. Soldiers in the rear endured shelling from far heavier 150-, 170-, and 210-millimeter batteries, as well as "a larger caliber gun nicknamed the Anzio Express."[67] The combined weight of these massive guns on the rear echelons was daunting, and for once the forward troops appreciated the relative safety that their position closer to the enemy afforded.

But for the most part, infantrymen felt enormous resentment toward friendly support units who were spared the carnage of the front. One rifleman serving in the ETO learned the "infantry taunt," a singsong chant aimed at the Army Service Forces, the Quartermaster and Transportation Corps troops who wore a blue-star shoulder patch. Infantry soldiers often teased those "Blue Star Commandos" with a derisive motto that equated "Service of Supply" with "Safe over Seas"; as one soldier recalled, the taunts were meant to rob the truck driver of his dignity by "saying that even his mother couldn't be proud of him."[68] And the rivalries cut both ways. During the Italian campaign, a column of paratroopers found itself the target of some infantrymen's gibes, disparaging the airborne soldiers' supposedly lighter combat burden ("You guys fight for a few days at a time and then you get relieved") and the extra $50 payment airborne units drew each month, claiming that paratroopers loafed most of the time and didn't fight much. The infantrymen characterized the paratroopers as "the glamour boys of the damned army." Far from friendly, such resentments could become bitterly serious. Looking back, the airborne soldier on the receiving end of the taunts could not quite understand how he had refrained from knocking out his antagonist's teeth with his rifle.[69]

Despite persistent resentments among different branches, the demands of modern mechanized combat created an enormously complex interdependence among disparate groups. Frontline troops begrudged support units, aircrew, and artillerymen their easier lives at the same time they depended upon the efforts of those soldiers to keep them alive. The expanded magnitude of the World War II battlefield required individual soldiers to place a great deal more trust in soldiers they could not see. The links between frontline soldiers and the artillerymen who supported them emphasized this brand of interdependence, all but unknown to soldiers of the War of Independence and the Civil War. In the mid-twentieth century, a GI's life depended in no small part on the actions of friendly soldiers he could not see and had never met. The tactics of the empty battlefield required riflemen to make these leaps of faith as a matter of course, as when they called in artillery support. That reality struck one combat soldier starkly as he watched an officer directing fire from a second-story window call in a strike on his own position, a building with German soldiers on the first floor. The officer yelled into the phone, "In thirty seconds fire concentration 52 and throw the book at it." Then the soldier watched as the officer dropped the phone, jumped out a back window, and sprinted away from the house, barely escaping before artillery fire leveled the building.[70] Such a strike testified to a number of changes in infantry combat: the flexibility of new communications technologies and the individual rifleman's dependence on the heavier shells of rearward batteries to attack dug-in positions. But it highlighted no change so much as the combat soldier's now-mandatory faith in comrades far outside the immediate sphere of battle and his trust in their competence, since even a small error in the aim or timing of the artillery blast would have cost the officer his life. GIs felt deep gratitude when their faith paid off and the infantry-artillery cooperation proved effective. During infantry training, infantrymen received instruction on the use of artillery in combat by experienced artillerymen. Though their teachers obviously brought their own biases to the lessons (the instructor "knew that we had been misled into thinking that infantry was the most important branch of the army"), the recruits left with a vivid message regarding the importance of the heavy guns. "When you get in trouble and hear artillery coming in," the instructor told the class, "it will sound like music. It will be sweeter than Glenn Miller's band." The students learned through experience the truth of the statement. A year later, in a hospital, the same soldier saw a news story from his division that described a line of tired, muddy infantry marching to the rear past its artillery support: "One rifleman stopped, looked at an artilleryman, and said simply, 'Thanks.'"[71]

But too often poor communication or simple incompetence turned the massive weight of friendly artillery against infantrymen. Slogging through the ETO, Harry Arnold's unit came under artillery fire. The unit had requested artillery support, and would have been delighted to see the artillery explosions, except that "most of it was falling in George company." Rumor within the unit held that the Americans' 2nd Division artillery was the source of the offending shells. The soldiers had no way of knowing whether or not the story happened to be true (Arnold himself thought that "the shell bursts had that howl peculiar to our artillery, and there was too much of it to be classed as a few shorts"), but the very fact of the rumor posed a devastating kind of threat to overall morale, giving the soldiers the suspicion that even their own military system was tilting the odds against them.[72] The presence of friendly air power presented another double-edged sword to infantry soldiers on the ground. In many cases, close air support poured devastating fire on enemy positions and appeared as a godsend to friendly infantry troops engaged on the ground. But when pilots failed to differentiate between their own troops and enemy troops, their presence posed a lethal and terrifying threat to riflemen.

Occasionally the presumed support turned into a threat in the blink of an eye. Soon after Harold Gordon's unit came ashore in Normandy, they witnessed a reassuring sight: a flight of British planes that cruised along overhead and then sprang into dives, "smoke trailing from their wing guns, while the faint, peculiar multiple 'rat-tat-tat' of aerial MGs drifted down." His comrades cheered the diving aircraft; it was, Gordon recalled, "an inspiring sight." Only later did the soldiers discover that the British planes actually had been strafing the Americans' own First Battalion accidentally.[73] The flexibility of modern weaponry introduced other new avenues of danger from friendly fire to the World War II infantryman. Timed-fuse explosives, left to detonate after a predetermined interval, for example, could not distinguish between friendly and enemy troops. A soldier at Omaha Beach learned that lesson in a particularly vivid manner: members of his squad had left such charges against a wall on a short detonation delay when another wave of troops landed on the beach. As soon as the landing craft gates opened, the new wave of soldiers—oblivious to the explosives about to go off—sheltered themselves behind the wall. A few "laid their heads on the demolitions beyond the obstacles for protection." In the confusion, the soldiers could not hear the warning shouts of their comrades, who had to physically pull the men off the wall before it exploded. They were only fifteen or twenty feet away when the charges went off; shrapnel from the wall struck one in the helmet.[74]

By the mid-twentieth century, American infantry soldiers found themselves in a dangerous predicament. In combat, comrades provided critical military support, but that dependence upon others lessened the degree of direct control the individual soldier perceived over his own battle outcome. Interactions with comrades in combat remained an essential component of GIs' willingness to fight on the battlefields of the Second World War, just as they had been for soldiers of the War of Independence and the Civil War. But the reasons for that importance had evolved fundamentally. Soldiers of the linear era experienced cohesion in its most literal sense; their successors on the empty battlefields of the mid-twentieth century rarely enjoyed the same tangible, physical cohesion. More and more, they experienced combat in a new way, paradoxically alone and yet not alone. Surveying the enormous changes occurring in ground combat at the close of the nineteenth century, Charles Ardant du Picq spoke directly to the necessary evolution in the nature of the relationships among combat soldiers in the new environment. Reflecting on the impossibility of maintaining the old literal cohesion of earlier battlefields in the face of daunting new weapons, he intuited that the bonds between soldiers in battle would nonetheless continue to be critically important in driving troops in combat. "It is a truth so clear as to be almost naïve," he held, "that if one does not wish bonds broken, he should make them elastic and thereby strengthen them."[75] As with so many of the changes that characterized individual soldiers' combat motivation on the empty battlefield, the creation of elastic bonds that persisted even amid the dispersion of modern tactics revolved in large part around emphasizing the infantryman's sense of control over his surroundings and, by extension, his chances for survival. Underlining the ways in which that survival depended upon cooperation with comrades strengthened those connections and gave soldiers practical reasons to maintain their solidarity even when new technologies forced them to disperse.

Conclusion

In May 1863, Samuel Byers fought in the battle at Champion's Hill, east of Vicksburg, Mississippi, part of the Union campaign to capture the Confederate stronghold. Years later, the day remained a vivid memory, one he described in graphic detail to give other young men who thought themselves eager for the adventure of war a glimpse of its grisly realities. Rather than a glorious display of individual heroism, the battle was a brutal ordeal for the infantry troops engaged on both sides. Byers recalled how the enemy soldiers appeared in the open, "a solid wall of men in gray," muskets at their shoulders, "blazing into our faces" with apocalyptic fury. The battle featured little maneuver on either side; instead, the two lines "stood still, and for over an hour we loaded our guns and killed each other as fast as we could." An artillery shell struck a nearby corporal directly in the face as he stood in the line. The day after the battle, one of Byers's fellow Federals visited his company in their camp and issued a chilling invitation: "*Go over to that hollow, and you will see hell.*" Byers and a few of his comrades went, and the scene they discovered there haunted them for weeks. Evidence of the previous day's fighting lay in a ravine, where Confederate troops retreating in haste had thrown hundreds of bodies into a shallow grave and covered them with a layer of earth. Rain had washed away the thin covering, and what Byers and his companions viewed there was bracing in its horror: hundreds of half-decayed corpses, some "grinning skeletons, some were headless, some armless." Wild dogs had ravaged some of the bodies, and Byers fumbled for words to describe the sight. "Dante himself," the veteran judged, "never conjured anything so horrible as the reality" that he and his fellows spied the day after the battle.[1]

Some eighty years later Harry Arnold experienced battle for the first time on a dark night in the European Theater during the Second World War. Like Byers, his descent into the chaos of combat affected him deeply. Battle, Arnold wrote afterward, had enveloped his platoon with the sights and sounds of "contrived mayhem" in "a brawl peopled by madmen." The din of battle — shouts in German and English, the persistent rattle of machine-gun fire, the roar of tank engines, and the howl of artillery, rockets, and bombs shrieked an assault on the ears. The flash of artillery and tank guns lit up the night sky. The final, eerie touch was the wailing of a siren on a German Tiger tank. To Arnold, that uncanny sound exemplified "the whole cataclysmic mess." Combat, he concluded, was nothing less than "Dante's inferno unglued."[2]

Together, the two descriptions touch on some of the most interesting continuities and changes that infantry soldiers experienced as battle became more technologically sophisticated from the eighteenth century to the twentieth. Some important commonalities wind through both accounts: the unavoidable evidence of battle's danger, the overwhelming sensory barrage and confusion of being under fire, and the grisly human toll emerge from both sketches. Too, there is the appeal to the Italian poet who chronicled his tour of hell itself to capture horrors that seemed to both participants to be almost beyond description. Read together, however, the two depictions also document some important evolutions in the individual soldier's experience of battle. While Byers fought shoulder-to-shoulder with his comrades in broad daylight against a wall of grey soldiers formed in a similar mass, Arnold found himself separated from his fellows on a darkened battlefield littered with the industrial wreckage of twentieth-century warfare. What combat demanded of each proved dramatically different. Similar in the broadest sense, the two soldiers' experiences were nonetheless distinct in the particulars.

For both men, the reality of fighting proved worse than they had anticipated. The chasm between expectation and reality was not uncommon. From the Continental soldiers of the late eighteenth century to the Civil War infantrymen of the mid-nineteenth century to the battered GIs who fought during the mid-twentieth century, nearly every participant admitted that actual combat proved far more terrifying and traumatic than they ever had imagined previously. Not every soldier found the strength to endure in the inferno of infantry battle. Some, like Garrett Watts in 1780, lost their nerve and fled when their lines broke. Others (like the numerous "coffee-coolers" who lingered behind the battle lines of Civil War armies) drifted to the rear whenever fighting seemed likely, discreetly absenting themselves from the trials of the front. Still others suffered a failure not of personal will but of physical capacity, paralyzed or otherwise unable to perform, victims of psychoneurotic breakdown who tried to fight but could not. Such cases became increasingly frequent over the three centuries of this study. Many of them were like the World War II GI whose ten days of fighting in the Ardennes ultimately rendered him physically ineffective as a combatant: after a week and a half, cut off by German tanks in the Bulge salient, hungry and in pain, he reported half a century later that he simply "went to pieces."[3] Infantry combat was an undeniably ferocious ordeal; the temptation to surrender to natural instincts and flee the source of the danger was acute. And yet the vast majority of soldiers in American infantry armies stayed and fought when called.

The factors that helped those American infantry soldiers face battle evolved over time. They had to, because as the nature of ground combat itself grew more complex, what battle demanded of foot soldiers in different eras changed so dramatically that motivators had to evolve as well. Many of the most important changes in the infantryman's experience of combat were rooted in the progressively more sophisticated technologies brought to bear on the battlefield and the tactical evolutions those new technologies necessitated. Armed with cumbersome, slow-loading weapons of limited accuracy, infantry soldiers of the eighteenth and nineteenth centuries deployed in tightly packed, dressed lines that helped compensate for the limitations of arms and communications technologies. In those close quarters, modeled behavior, mechanical tasks overlearned through repetition, coercive threats, and the physical reassurance of comrades combined to help soldiers endure the savage butchery and perform amid the terror and gore of the firing line. In the twentieth century, quicker-firing, more accurate arms rendered the tightly packed ranks of the eighteenth- and nineteenth-century battlefield suicidal. To limit their exposure to the lethal "storm of steel" that those new weapons unleashed, soldiers dispersed from their linear formations and employed cover, concealment, and move-and-fire techniques that restored their ability to take the tactical offensive. But at the same time those dispersed tactics restored troops' ability to operate on the battlefield, they eroded the influence of the traditional factors used to drive troops to face fire. Mutual surveillance, direct coercive threats, and the tangible physical reassurance of comrades lost much of their power to motivate as troops spread out in combat. Dispersed tactics forced soldiers to muster the will to face enemy fire without the certainty of reprisal from officers if their nerve failed; dispersed tactics also frequently denied combatants the physical touch of comrades. At the same time, the heightened lethality of the industrial battlefield made it far more difficult to make reasonable soldiers willing to face combat's dangers; the exponential increase in deadly projectiles threatened to make even brief exposure to battle seem a death sentence. Recognizing that truth in his comparative analysis of German and American combat effectiveness during the Second World War, Martin van Creveld wondered how (as the technology of warfare became ever more destructive) "the balance between rationality and irrationality" could be struck.[4] It remains a supremely difficult question—but it is a question that must have answers, since so many soldiers in different centuries have managed to face enemy fire and participate in battle with some degree of effectiveness. Those veterans offer ample evidence that the persistent questions of motivating troops in combat were indeed solved to some degree.

It is equally apparent that the simple and enduring question "What makes soldiers fight?" has no similarly straightforward answer. In his synthetic work on combat motivation, social scientist Anthony Kellett suggested nearly thirty years ago that, in the main, soldiers continued to fight (if sometimes reluctantly) simply because the "penalties of not fighting" overwhelmed "the uncertain risks of fighting itself."[5] That assertion anchors some interesting observations about the relationship between soldiers' motivation for combat and technological change. Certainly many of the factors that motivated infantry troops in battle operated in both the linear and dispersed systems, since so many of the things that battle demanded of soldiers in both eras (stifling natural instincts and hazarding exposure to the violence and confusion of battle) remained so similar. What changed over time as a result of technological and tactical evolutions was the mixture and application of various motivators: a shift from emphasizing the obvious penalties of not fighting to focusing on the uncertain risks of fighting itself. To compensate for the weakened reliability of threats against soldiers who refused to fight (the growing difficulty in relying on physical and social coercion with effect, for example, on the empty battlefield), the factors that drove soldiers on the dispersed battlefield gradually shifted the focus from promising to directly punish defection to shading the soldier's imperfect understanding of the risks of fighting itself: specifically, by suggesting that particular behaviors could reduce those risks dramatically. That subtle interpretation helped bolster a belief that committed participation in battle was not necessarily a death sentence by suggesting that even amid the maelstrom of industrial ground combat, specific decisions—the individual soldier's own agency—could help increase the odds of survival.

The importance of that increasing emphasis on the individual soldier's agency on the dispersed battlefield played out in a number of ways as combat became more depersonalized, more isolating, more specific, and more continuous. It helped explain the transition from training that focused on rote drill and mastering a handful of straightforward tasks through overlearning to regimens that emphasized realistic "battle inoculation," offering soldiers experience with the physical sensations of fear and presenting battle as an environment filled with choices in which specific decisions had markedly different influences on the chances of survival. The emphasis on conditioning soldiers of the mid-twentieth century to view the battlefield as a series of consequential choices for individual foot soldiers also helped explain the transition from a system that privileged conspicuous displays of physical courage from combat leaders to one that placed new importance on competence and proved (by earlier standards) surprisingly forgiving of lapses in the stoic facade. Soldiers' changing relationship with their weaponry put more decision-making power in the hands of the individual even as the nature of infantry combat rendered that relationship more complex and problematic. By the mid-twentieth century, the soldier's arms were no longer simply the way to respond to enemy threats but also represented a tool whose very use might place the bearer's own life in jeopardy, threatening to reveal a hidden position to opponents or even exploding in the operator's hands. And the new focus on agency also suggested a way to look at the cohesive bonds between soldiers, using the pragmatic desire to stay alive as a prism to understand how interactions with friendly soldiers evolved over time to further that urgent goal.

Rather than present participation in battle as a fundamentally irrational act forced on rational men, attention to individual perceptions and decisions provides some insight into the ways that committed fighting even on the modern industrial battlefield could be reframed as a series of rational, if highly unpleasant, choices. Near the end of the nineteenth century, Charles Ardant du Picq lamented the daunting necessity of motivating soldiers on the industrial battlefield, where technological innovations removed enemies to greater distances and created a battle space that was simply too lethal to bear in the exposed ranks of traditional infantry formations. He acknowledged the difficulty of motivating soldiers in such an environment, where the contest appeared to be not with a human enemy but with fate itself. The changing nature of relationships with training, leaders, weapons, and comrades presented here suggests ways in which the experience of combat was reframed for individual participants until it no longer appeared as a struggle against fate but as a series of choices that could be mastered to some degree. That argument echoes the second of Paul Fussell's three phases of awareness in battle. After the individual shed his initial, grossly unrealistic notion that he was simply too special to die in combat, Fussell held that most settled on a slightly more realistic belief: "It can happen to me, and I'd better be more careful." In that second stage, soldiers believed that they could avoid the danger and death of battle by keeping alert at all times, being better trained than the enemy, exercising care when firing a weapon so as not to reveal a position, and employing a host of other factors—all of which fell under the soldier's control.[6] That belief, in turn, helped soldiers face the daunting rigors and demands of industrial combat on the dispersed battlefield.

A case specific to the Second World War complicates that connection between individual soldiers' perceptions of their own agency and their ability to face battle and suggests one area for further research. Given the nature of their duties in combat, medics do not easily fit into the framework. Medics hazarded the danger and trauma of the battlefield, just as the infantry soldiers discussed in the previous pages. Unlike those combatants, however, they lacked the same opportunities to positively influence their outcomes in battle. Preoccupied with their duties, medics had precious few opportunities to respond constructively to the deadly threats leveled against them; the very nature of their duties required them to focus on serving others in battle, rather than acting in ways that might improve their own odds. Yet American medics performed remarkably in combat, acting selflessly even while exposed to the dangers of battle. As a group, they offer an opportunity to further test the connection between perceptions of agency and the willingness to face the rigors of the empty battlefield. Understanding how members of this cohort differed from their counterparts in rifle companies (how their selection, training, and preparation, for example, contrasted with similar regimens in the infantry, and how those factors affected the way they perceived their time under fire and responded to its strains) presents a new way to approach the argument in these pages—as well as offering an opportunity to examine the effects of altruism on behavior in battle.

More broadly, the study of these evolutions in the experience of ground combat from the end of the eighteenth century to the middle of the twentieth offers some ways to think about the manner in which infantry soldiers have encountered battle in the first years of the twenty-first century. In the last fifty years, American soldiers have fought in a number of large-scale conflicts: the war in Vietnam, the 1991 Gulf War, and, more recently, lengthy campaigns in Iraq and Afghanistan. Evidence from those battlefields reveals some intriguing developments in the nature of infantry soldiers' experiences in battle, as well as some continuities that reach back through the centuries. Certainly some of the most fundamental elements of infantry combat emerge just as clearly from accounts of more recent battlefields: the inevitability of death and destruction, the overwhelming confusion of battle, the all-important necessity of stanching visceral fears to perform amid the chaos. And the overriding desire to survive appears prominently in those musings: like their predecessors, the American soldiers fighting in the wars of the twenty-first century desperately want to stay alive, and their experiences in battle often clarify those feelings. A twenty-one-year-old lance corporal in the 2003 invasion of Iraq described looking on a dead body in battle and the stark way that sight focused his own concerns: "Everything in life is overrated except death. All that shit goes out the window—college, nice cars, pussy. I just don't want to end up looking like that dude who looks like a box of smashed tomatoes."[7]

Other commonalities connect these more recent combat infantry soldiers with their predecessors in earlier armies. Since the Second World War there has been no tactical revolution analogous to the one necessitated by the advent of the storm of steel at the beginning of the twentieth century. Dispersed tactics still reign on the battlefield; indeed, they remain virtually the only way for a military force to weather the omnipresent danger modern weaponry presents in order to operate effectively against an opponent. Thus many of the trappings of recent battlefields bear more than a passing resemblance to the fields of the mid-twentieth century. Because stealth remains vital to survival, familiar features of the World War II GI's experience, like silent hand signals (used, as a veteran of the 1991 Gulf War recalled, "to slow or increase the speed of the patrol, to call team attention to a particular area in the patrol zone, to call certain members forward for conference, to initiate a firefight and to cease fire, among other actions," especially in situations in which yells might give away a hidden position, or when the din of battle overwhelmed the volume of shouting), have become even more important on the battlefield. And soldiers' common response to the chaos of combat shares a timeless quality with generations of their predecessors: "Sometimes, because of battlefield stressors such as exploding enemy ordnance, marines forget what the hand signals mean, and the situation becomes loud and confusing and marines begin yelling rather than using their hands."⁸ Nighttime offers modern soldiers no respite from the dangers of combat; indeed, since the American army's twenty-first century night-vision technology gives its foot soldiers an advantage in darkness that none of its opponents will match for the foreseeable future, night fighting has become one of the hallmarks of the American military's conventional operations. Generations of soldiers dating back to the War of Independence have benefited from the effects of overlearned tasks amid the panic of battle, manipulations conditioned through endless repetition in drill to become almost second nature; while the equipment available to American soldiers on the battlefields of the twenty-first century is far more technologically sophisticated than Continental soldiers or their successors in the armies of the Civil War could have imagined, the sensations produced in battle by such focused practice remain familiar. One marine sergeant who survived a harrowing Humvee ride through an Iraqi town whose main thoroughfares militants had converted into a gauntlet of automatic fire confessed that he felt nothing in the thick of the firefight: "It was just like training. I just loaded and fired my weapons from muscle memory. I wasn't even aware what my hands were doing."⁹

Some of the trends outlined in the first chapter—the characteristic changes in the individual's experience of combat that accompanied the transformation of the battlefield as linear tactics gave way to dispersed tactics—have continued in recent years; some have accelerated. The depersonalized nature of battle continues to grow, and the sense of isolation that dispersion creates visits the foot soldiers of the twenty-first century equally often. If anything, longer-range weaponry, better camouflage, and more frequent night operations have made the sense of removal from both enemies and comrades more acute at times. (The army's continued emphasis on keeping soldiers connected in other ways when tactics no longer permit them to cohere physically—providing soldiers microphones and earpieces in their helmets, for example, allowing them to remain in constant communication even when spread out—exemplifies one way that technology attempts to address the detrimental effects of isolation on the battlefield.) Indeed, soldiers since the Second World War have faced new, increasingly sophisticated technologies that render the depersonalized postindustrial battlefield even more frustrating. Troops on patrol in Vietnam, for example, quickly learned to be terrified of the booby traps and mines their enemies scattered for them, searching the underbrush for trip wires and listening keenly for the telltale *click* that accompanied a misstep that activated the triggering mechanism. Soldiers deployed in Iraq during the insurgency fear the same kinds of traps (improvised explosive devices, or IEDs, have been the cause of most American deaths both there and in Afghanistan), with a menacing twist. No longer simply passive bombs with automatic mechanical triggers, many are more advanced contraptions that can be actively detonated remotely by human operators, introducing a new and vexing asymmetry for American soldiers in battle. Targets of such devices must weather not just the danger of explosions but the constant suspicion that someone is watching and timing their movements and that even the most innocuous-seeming debris might camouflage a deadly package. A veteran of the Iraq conflict noted the toll such weapons took on his comrades: "Many of them were killed by vehicle-borne IEDs. The bad guys put them in vehicles or just plant them on the side of the road, and as we're going by, they press the button and it explodes."[10] Certainly the sense that combat has become more specific persists; indeed, that feeling appears heightened by some of the technologies that define more recent battlefields. Those sophisticated IEDs, the iconic weapon of the campaigns in Iraq and Afghanistan, combine the frustration of a weapon without an operator who can be retaliated against with the terrifying sense of having been personally singled out as the target. The early, crude versions of the devices (triggered mechanically by the victim) closely resembled the mines of the Second World War. The more advanced IEDs, triggered remotely by an unseen enemy (either hidden out of sight or blending in among the population) who monitors the approach of targets, thus combine two of the most psychologically corrosive elements of industrial warfare.

Other elements of ground combat have changed even more dramatically. With some limited exceptions (a few battles in Vietnam, the ninety-six hours of ground fighting during the 1991 Gulf War, and the first weeks of the 2003 campaign in Iraq), most American infantry soldiers have not faced the large conventional battles against uniformed enemies employing their own armor, artillery, and air support that were so familiar to the GIs of World War II. Increasingly, American troops have found themselves facing a new kind of opponent on a new kind of battlefield: enemies who frequently wear no uniform and prove almost impossible to distinguish from noncombatants, and who often force battle in populated areas. The massive clashes of peer armies that characterized the Second World War have given way to more frequent irregular, asymmetrical engagements: "war amongst the people," in the phrase of British General Rupert Smith.¹¹ That shift has affected countless elements of U.S. military planning, policy, and doctrine; it has also dramatically changed the way individual soldiers experience the rigors of combat.

If anything, warfare in populated areas against an unconventional enemy demands an even higher degree of autonomy and initiative from individual infantry soldiers than did combat in the Second World War. Asymmetric, postindustrial warfare presents troops with a staggering number of life-and-death reactions to make; at the same time, the very nature of irregular warfare (intentionally obscuring important information about combatants and the battle environment) makes the choices facing the individual soldier even more confusing, the options and outcomes unclear. Does that pile of trash hide a booby trap? Is the driver of the vehicle racing toward the checkpoint a suicide bomber or merely a terrified and confused civilian? Is the figure running away from the building a member of a militant group or simply a frightened schoolboy? (A sniper with the First Infantry Division captured those frustrations in his description of evaluating potential targets at night: "You can't quite tell what someone is holding. There's a lot of argument between snipers while they're watching. There are three guys. One has a regular-vision telescope. The other person has thermographic vision. The other person is looking through a magnifying scope in the rifle. Everybody is arguing." Those arguments revolved around the impossibility of pinpointing objects and intentions: "What has he got? Is it a gun? No, I don't think it's a gun. Is it a bomb? It's a bomb. No, it's an IED. No, it's a toolbox. It's a toolbox, yeah, it's a toolbox. So he doesn't die." Of course, other figures present very real threats: "Sometimes, Yeah, he's got a shovel. He's digging. He's digging on the side of the road. He's going back to the truck. He's got something in his hand. He's going to bury it. Take him out. Bang.")¹²

Combatants confronted with such scenarios possess enormous flexibility of action, but must rely upon imperfect and incomplete information when making crucial decisions. As the pace of battle escalated, so too has the urgent necessity of making split-second judgments. Given the ever-growing destructive capacity of portable weaponry and the frequent need to utilize those tools in areas populated with noncombatants as well as with legitimate targets, the costs associated with erroneous decisions have grown as well. Together, those phenomena have increased both the spectrum of possible actions available to individual soldiers in a given situation and the stakes attached to wrong choices. The corresponding pressure on individual soldiers and their judgment appears even more pronounced than in previous incarnations of the dispersed battlefield. The new battlefield is itself different: controlled by American troops but populated by innocent people who cannot and must not be harmed, both by law and in order to win the war. The presence of those noncombatants, and the near impossibility of distinguishing them from legitimate targets, aggravates the already staggering level of stress to which troops are exposed: a soldier might unintentionally commit a war crime while defending him- or herself. It is not yet clear how prolonged exposure to such new conditions of battle—the prevalence of threats that exact a brutal toll on soldiers in combat but which can only be avoided or neutralized, not directly attacked themselves—will affect individual soldiers. Reports from both twenty-first-century theaters suggest that the incidence of stress- and trauma-related psychological disorders may be higher than one soldier in three; among other things, such high rates of stress disorders may indicate the kind of damage that such sporadic and one-sided battle inflicts on individual psyches. The pervasive sense of being targeted by a distant, unanswerable enemy can be so overwhelming and so psychologically punishing that the recipients of other kinds of attacks must actively assert an objective, intellectual understanding as a bulwark against more immediate, reflexive responses. One embedded reporter in Iraq reminded himself during a bombardment that "mortars fall in a totally random pattern. It's not like there's a guy crouched somewhere in a field with a rifle, trying to pick you up in his scope. You're not being individually targeted." That enforced awareness was meant to offer some psychological protection while enduring life-threatening but unanswerable attacks: "You have to take comfort in the randomness of it all."[13]

The trend in the duration of battle, on the other hand, has not necessarily continued toward longer exposure to danger. Continental soldiers weathered combat a half-hour at a time, perhaps every month or two; soldiers of the Civil War fought for hours (and occasionally days) at a time, every few weeks. During World War II, frontline GIs often found themselves under fire or the threat of fire for days or weeks without a break, and enjoyed no respite from danger at night or during foul weather. American infantry soldiers of the past half-century have experienced combat as a more episodic—and often more unpredictable—event that lacks the defined front line of earlier conflicts. From the Vietnam soldiers ferried via helicopter into battles that lasted days at a time before withdrawing to more secure and relatively comfortable bases, to the soldiers of the 1991 Gulf War who achieved their objectives in four days of fighting after months of tense buildup and waiting in the desert, to twenty-first-century soldiers in Iraq who use the heavily fortified and heavily defended Green Zone as a haven, the formless wars of recent decades have furnished an unusual and frequently unnerving mix of safety and danger. American combat soldiers of the early twenty-first century have at least one other trait in common with their predecessors of the eighteenth, nineteenth, and mid-twentieth centuries: they continue to fight when called to, even in dangerous, difficult, and frustrating circumstances. Nineteenth-century theorist Jan Bloch's assertion—that industrial warfare would become too terrible for any reasonable person to face—seems no more true now than when he wrote it a hundred years ago on the eve of the World Wars. The twenty-first-century campaigns in Iraq and Afghanistan (not to mention the war in Vietnam and the first Gulf War) provide robust evidence that, in the main, American soldiers have met the trials of postindustrial, asymmetric warfare, just as their predecessors in earlier wars met the particular challenges of those conflicts. In some regards, the precise mechanisms that help the soldiers of the early twenty-first century master their fears and summon the will to face danger remain mysterious. Soldiers and scholars studying these conflicts continue to invoke unit cohesion to explain many of the risky, selfless, and sometimes inscrutable behaviors witnessed in battle. The U.S. Army's own 2003 study *Why They Fight: Combat Motivation in the Iraq War* concluded that such emotional bonds remained "a critical factor in combat motivation." American soldiers in Iraq, the study's authors argued, "continue to fight because of the bonds of trust between soldiers."[14] That analysis rested on interviews with combat veterans, soldiers who agreed that "we were just fighting for one another," or that "it was just looking out for one another. We weren't fighting for anybody else but ourselves."[15] The same sentiment has since appeared in scores of anecdotes and offhand remarks from veterans of the fighting in those conflicts. For one marine, that was a basic, unavoidable truth of war: "All you think of is your friends, the men you are fighting with. It is self-sacrifice so that the guys around you will have it just a little bit easier, and they're doing the same to try to make your time a little bit better." Another soldier, watching rounds illuminating the sky during the 2004 battle at Ramadi, reported that the fate of his comrades consumed his thoughts: "All I could think of were my friends and if they were all right, if they were all going to come back or not."[16] Another marine characterized the cohesive bonds among the members of his unit as so profound that he and his comrades rarely needed to discuss them aloud: "No matter what happened and no matter what the circumstances, we would look after our comrades before looking after ourselves. Sometimes we talked about this concept, but most of the time it went unsaid; we just knew what we'd do for one another. If a man went down, another would be there to drag him away or to cover his body with his own."[17]

The preceding chapters outlined some of the weaknesses in the unit cohesion explanation for combat motivation. One of this book's central arguments is that a sense of controllability (the belief that individuals could improve their odds of survival by behaving in specific ways) proved more critical to the motivations of World War II GIs on the dispersed battlefield than those emotional bonds as traditionally understood. A sense of controllability proved critical to soldiers' combat effectiveness when it aligned their desire to stay alive with militarily useful actions: spreading out, digging in, keeping close to the opponents' infantry so as to minimize the enemy's ability to employ mortars and artillery, and dozens of other small but consequential behaviors. That framework provides one way to think about the ways that soldiers of the twenty-first century have been prepared to face battle. Indeed, emphasis on the controllability of battle and on the individual's own agency in combat appear to have grown more pronounced among foot soldiers of recent decades. Anthony Swofford, a Recon Marine who fought in the 1991 Gulf War, summarized the thought process of an elite marksman evaluating his failure in the excruciatingly difficult test of placing a series of shots so close together on a target that the resulting holes could be covered with a ten-cent coin (a "dime group") at a distance of 1,000 meters; the explanations all revolved around the shooter's own personal failings. "There are reasons you're not hitting a dime group at a grand. . . . You hadn't completely expelled your breath when you shot. You are afraid of the rifle. Your spotter gave you the correct dope but you dialed the scope incorrectly. You are tired. You are stupid. You are bored. You are a bad shot. You drank the night before. You drank excessively the night before." The brutal cause-and-effect of the battlefield drummed into these elite soldiers proved spectacularly unforgiving: "These are all unacceptable reasons for not achieving a dime group at a grand. A nickel group is occasionally acceptable. A quarter group and you are dead." The supremely predictable battlefield they confronted afforded no margin for error: "You have missed the target but the target hasn't missed you. You must remember that you are always a target. Someone wants to kill you and their reasons are as sound as yours are for killing them." The exhaustive list of reasons an individual might fail to achieve a dime group lays out a daunting catalog of things that can go wrong, with deadly consequences; placing the responsibility for those failings squarely on the shooter himself simultaneously offers the tantalizing reassurance that potentially lethal missteps can be conquered through determined mastery. That construction of battle is unforgiving. Snipers who are careless, untrained, lazy, or untalented suffer a straightforward penalty: "Your enemy will be the last person to witness you as a living thing. He'll acquire you through his optics and he will not pause before pulling the trigger."[18] At the same time it offers comfort in the control it affords the individual over the outcome: kill the enemy first, through superior talent and preparation, and the shooter survives.

As it had for soldiers of the mid-twentieth century, that framing of the battlefield as an environment where the laws of cause and effect still held requires careful conditioning. Subtle lessons repeated over and over in training emphasize the consequences of individual choices and actions, however small. One marine registered that lesson during his training for one of the elite sniper units: as the candidates began a road run, one of the sergeants barked, "I don't want to hear anyone's boots hitting the ground. That's a target indicator and it'll give your position away to the enemy!" (The simple reminder upended some previous understandings of combat behavior. Thirteen-man marine squads treated noise from footfalls as unremarkable; in the elite two- and four-man sniper teams, such noise could alert a numerically superior enemy to their presence, with potentially devastating consequences.)[19] The overt message: *be as quiet as you can at all times, even when moving.* The more subtle lesson: *what you do or fail to do in battle helps decide whether you live or die.* As proved the case with World War II GIs, those lessons find an eager audience with many contemporary ground troops precisely because they imply a correlation between ability and survival and suggest that the individual exerts some control over outcomes. Some soldiers naturally reinforce those observations on their own: in a letter that echoed one in which a World War II soldier assured his parents that he would survive combat because he would avoid careless mistakes, Antonio Espera wrote his wife from Iraq in 2003 that "I've learned there are two types of people in Iraq, those who are very good and those who are dead." His confident appraisal of his own chances: "I'm very good."[20]

Of course, none of these twenty-first-century soldiers enter combat as blank slates. Like their predecessors in earlier armies, they come to battle with expectations regarding the nature of the experience and what it will demand of them. Just as the soldiers who fought in the Second World War looked to previous generations for insights into the nature of combat, and troops who fought in Vietnam took some of their cues from World War II–era movies and their relatives' stories of service in the ETO and the Pacific, combatants in the 1991 Gulf War in turn drew on a rich and varied body of Vietnam films to inform their impressions of battle. Soldiers deployed in the 2002 and 2003 invasions of Afghanistan and Iraq similarly formed impressions of battle from modern war movies. Some, like *Saving Private Ryan*, celebrate the exploits of their grandparents' and great-grandparents' generation; others, like *Black Hawk Down*, portray the chaotic realities of modern combat. As the first chapter noted, veteran soldiers frequently marveled at how different their actual experiences of combat proved relative to their preexisting impressions. (Unlike their predecessors, however, the new generation of soldiers have not just provided a merely passive audience for images of combat; many formed a separate set of expectations about combat influenced through active engagement with first-person military-themed video games, some featuring shockingly graphic and realistic images. From simulations that digitally recreate famous battles from the Second World War to the U.S. Army's own first-person shooter simulation, many of those serving in the American army of the early twenty-first century bring to their experience preconceived notions of battle shaped not just by passively absorbing film images but by actively participating in virtual combat—games that dramatically exaggerate the controllability of the combat environment.) In addition, many soldiers in the 1991 Gulf War and the invasions of Afghanistan and Iraq trained at the U.S. military's hyperrealistic National Training Center at Fort Irwin, California, which employs a bevy of advanced technologies that enable ground troops to rehearse fighting maneuvers with a degree of sophistication far beyond that offered by the Second World War's crude "battle inoculation" exercises. How these new forms of conditioning have helped shape soldiers' perceptions and motivations in actual battle warrants further examination.

One other fact remains relatively certain: soldiers will seek control, or the sense of control, even (or especially) when the nature of combat denies it to them. The leader of a marine platoon in Iraq reflected on both the feeling of powerlessness created by IEDs and guerrilla tactics and the counterbalancing sense of efficacy created by offensive actions against real opponents: "After a month of walking around Ramadi feeling as if we were more or less unsuspecting targets, it felt good to hit back strongly, to regain some of the initiative, to kill our enemies."²¹ Nor does the sense of control need to be objectively realistic. Newly promoted marine scout/snipers, for example, receive the honorific "HOG" ("Hunter of Gunmen," indicating that the graduate has completed the most demanding training the Corps offers and is now qualified to stalk any target, including the enemy's own snipers) and a 7.62mm bullet, known as a "HOG's tooth," and worn as a necklace at all times. The bullet's significance is simple: "It is a charm, meaning that it is the only bullet meant for that marine." According to a ritual familiar to World War II GIs (many of whom had, some seventy years earlier, fastidiously guarded the slug from a particularly close near-miss as "the one with your name on it"), possession of that special bullet renders the owner impervious: "In combat, no other sniper will have a bullet for him."²²

In the mid-twentieth century, the connection between following instructions and increased chances of survival was frequently implicit: that was the point of the phony "Brown the Clown" tombstone, carefully orchestrated live-fire exercises in which no one who followed instructions carefully suffered harm, and the parades downrange to witness the devastating effects a small mortar team could unleash. Hundreds of subtle cues conditioned soldiers with reassuring (and militarily beneficial) lessons: *There are things you can do to save your life. Do it like this and you'll be okay.* In the all-volunteer forces of the early twenty-first century, the connections between the level of military skill, a willingness to follow instructions precisely, and the soldier's own survival often appear not as an implicit suggestion but rather an explicit promise. An infantry sergeant who served multiple deployments in Iraq following the 2003 invasion confided to a reporter that he made the connection unambiguous to his charges: "When you train these boys, you tell them every day, 'You do what I tell you, and I will get you home.'"²³

Given the unalterably capricious nature of combat, of course, no one can make good on such a guarantee. Like earlier foot soldiers, modern combatants often try to dodge this awareness, especially at first. After his first weeks in combat in Iraq, a platoon leader noted that he had not truly assimilated the idea of combat deaths in his platoon: "There was no real reason underlying my avoidance of reality, no reason other than that death hadn't yet happened to me, to my platoon, to Joker One." That view was, he acknowledged, not solely a function of the platoon's ability: "There wasn't any particular skill or fortitude in this—we had simply been lucky." At first, the platoon leader attempted to replace the reality with an encouraging sense of control. "But I didn't know this at the time," he observed. "Instead, I reasoned that if we had made it through the fierce fighting of April unscathed, then we could, and probably would, make it through anything and everything that the rest of the deployment would throw at us."[24] The nature of the new weapons and new tactics these soldiers face, however, supplies unwelcome reminders that luck determines as many outcomes as skill. Marine Lance Corporal Ryan Mathison demonstrated that truth in the most dramatic and terrifying way possible: while on patrol in Afghanistan in 2010, he accidentally placed his boot on the trigger of a hidden IED. Mathison's weight activated one of the booby trap's two blasting caps, which exploded with a pop and tossed stones into the air. Inexplicably, however, the trigger failed to ignite its payload, described as "a powerful mix of Eastern Bloc mortar rounds and homemade explosives spiked with motorcycle parts, rusty spark plugs and jagged chunks of steel." The response of Mathison's fellow Marines—at least one of whom was close enough that he likely would have been killed in the blast as well—blended awed disbelief ("Goddamn Matty, man") and the inescapable lesson of the incident: "Lucky son of a bitch." While waiting for a specialized team to neutralize the explosive, Mathison himself first downplayed the magnitude of his near miss ("I'm still calling it nothing. I'm going with that it was nothing. . . . Makes me feel better") as his squadmates eagerly connected his seemingly miraculous escape with any number of fluky coincidences, including the fact that one of their fellows had uncharacteristically read from the Bible the previous evening. ("I saved you," the Bible-reading comrade claimed.) As his fellows around him reflected on the number of free beers his brush with death would merit when they returned to their base, Mathison evinced a desire to make some changes in his life—and perhaps to reassert some small sense of control over his own destiny: "If this really was an I.E.D., then you ain't drinking with me. Because I'm done drinking. I'm going back to the way I was before I joined the Corps."[25] Most infantry soldiers ultimately face some evidence that their own ability to influence their fates pales in comparison to the mountain of factors that lay beyond the control of any individual. One of Swofford's comrades found a degree of solace if not comfort in his realization of both the power of the individual soldier's agency and its limitations: telling Swofford to "cool down" after an outburst, he reminded him that the foot soldier "can't control anything but our crosshairs."[26]

How the cumulative effect of these evolutions and revolutions in the nature of warfare have affected the experience of fighting from the perspective of the individual ground soldier is a question that is just beginning to be answered. And the commonalities and differences in the ways these more recent American infantry troops—since the institution of the All-Volunteer Force in 1973 a smaller, more professional force than that which fought in the conflicts of earlier centuries—have been motivated from within and without to face the trauma of battle is an intriguing and important question in its own right. Given the magnitude of the changes in the character of battle observed as conventional industrial combat gave way to asymmetric, unconventional, irregular warfare from the Vietnam war to the twenty-first century, the examination of the evolutions in the experiences of ground combat (and soldiers' motivations to endure its dangers) is properly the subject of its own book.

For the GIs of the Second World War, recasting battle as an environment subject to some degree of individual control helped reduce soldiers' undermining sense that industrial combat was, at bottom, an arbitrary and deadly meat-grinder. Heightened perceptions of control over the outcome, in turn, helped increase soldiers' willingness to face battle. But that construction could not hold indefinitely. Exposed to the acute rigors of twentieth-century combat, soldiers increasingly came to believe that everyone had a limit on the amount of stress that could be endured. As one of Samuel Stouffer's interviewees put it, "I believe through experience that a man who has seen two campaigns shouldn't see any more action." Given enough exposure, he argued, "the horrors of war will get any man down." After two campaigns, the veteran assured his interviewer that he and the men of his unit had all reached their breaking point, though their bodies remained physically capable of fighting. "Take it from me, a voice of experience," he warned. "If my company makes one more invasion you had better tell the medical corps to be sure and have 42 straight jackets as there are only 42 of us left."[27] The common World War II adage "Every man has his limit" had an empirical basis: studies conducted among American GIs during the Second World War found that, in general, troops could tolerate a fixed number of days of battle. More than 200 or 250 days in combat, it seemed, and few soldiers could bear the terror and stress of battle. The incidence of psychoneurotic breakdown—"combat fatigue," in the parlance of that war—increased sharply. Past a certain point, chronic and repeated overactivation of the body's alert systems simply ground down the psyche until all but the most exceptional could no longer function.[28] Here too the experiences of World War II soldiers appear not altogether unique. Soldiers of the Continental Army and the Civil War armies were far less likely to see that amount of combat; the sporadic nature of fighting in the eighteenth and nineteenth centuries meant that few soldiers spent two hundred days under fire. While in the twentieth century the phenomenon of psychological breakdown in soldiers became linked to the particular experience of industrial combat, careful examination of experiences from earlier wars has provided persuasive evidence that such psychological breakdown surfaced in every conflict. Historian Eric Dean's comparative study of veterans from the Civil War and the conflict in Vietnam suggested that some mid-nineteenth-century soldiers suffered from symptoms akin to those that twentieth-century psychologists termed post-traumatic stress disorder: sleeplessness, paranoia, unpredictable bouts of rage, and lifelike flashbacks.[29] The similarity of symptoms suffered by both nineteenth-century Civil War soldiers and the twentieth-century veterans of very different kinds of combat (in many cases, the physical manifestations of the trauma appeared nearly identical) suggests another important continuity in the way that individuals responded to the stress of battle, regardless of the specifics of the experience.

That continuity, in turn, recalls the third and final stage of realization and understanding that Paul Fussell described in combat soldiers—the only one of the three that reflected "accurate perception" rather than hopeful rationalization. Given enough exposure to combat, Fussell argued that the belief that talent and soldierly ability could keep a soldier alive ultimately gave way to "the perception that death and injury are matters more of luck than skill." That acknowledgment produced an inevitable third realization: "It is going to happen to me, and only my not being there is going to prevent it."[30] The failure of combat to operate according to soldiers' rules and expectations proved acutely debilitating. Soldiers conditioned to believe that they could exert meaningful influence over their personal outcomes might experience a boost in alertness and morale upon first experiencing combat, as the desire to be faster, sharper, and more effective than an opponent provided a surge in awareness and attention. But despite persistent attempts to convince them that combat would in fact present such a controllable environment, soldiers found that faith shaken and ultimately destroyed by the unavoidable evidence of battle's randomness. Who was hit and who was spared obeyed no predictable or controllable set of rules: being a capable and careful soldier was certainly no guarantee of survival. As Harry Arnold, fortunate enough to survive his slog through northwest Europe during the Second World War, put it, "No survivor, unless a fool, attributes his good fortune entirely to his own machinations. Time after time we survived because of what someone else did, wittingly or otherwise."[31] Such a summary reflected a veteran soldier's admission that the battlefield was ultimately a place beyond individual control; in Arnold's case, that belief was replaced by a much more realistic view of the capricious nature of combat.

Perhaps what is most striking about the evolution of the infantryman's experience over these two centuries is that so many soldiers endured the inferno of battle for so long. That the army could take ordinary Americans and prepare them to withstand the dreadfulness of infantry combat in the radically different circumstances of three different wars indicates a tremendously flexible system that continues to adapt to new demands—even demands that initially appear all but insurmountable. The ability of the military system to prepare individuals to withstand trauma has progressed sufficiently to absorb the changing trials of industrializing combat.

But enduring the trauma of battle came at a price, and it changed the participants. Confronting the gap between expectation and experience, and repeatedly placing the natural instinct for preservation in direct conflict with the demands of the military system in combat, changed the way soldiers approached and reflected upon their experiences. Desire for glory and medals often faded as one objective—survival—crystallized in the minds of the participants. That urgent and overwhelming goal crowded out others. Reflecting on the longing, common among many soldiers of the Second World War, to bring home a German Luger or enemy flag as a trophy, Elmer Miller, a Marine veteran of some of the harshest fighting in the Pacific, admitted years after the war that he had given little thought to bringing home prizes from his battlefields. "You know, a lot of guys like souvenirs," he told the son of one of his former officers. "I got a good souvenir. I got my life." Through tears, Miller confessed, "That's the only souvenir I ever wanted. That's the best one you could get."[32]

Notes

Introduction

1. David Thompson in Robert U. Johnson and C. C. Buel, eds., *Battles and Leaders of the Civil War* (New York: Century, 1882), 2:662.

2. Roy R. Grinker and John P. Spiegel, *Men under Stress* (Philadelphia: Blakiston, 1945), 38.

3. Historian John Lynn termed the factors that enabled soldiers to face the trauma of battle "combat motivation"—distinct from "initial motivation" (the factors that led individuals to join the army in the first place) and "sustaining motivation" (the factors that kept soldiers with the army on campaign). John Lynn, *The Bayonets of the Republic: Motivation and Tactics in the Army of Revolutionary France, 1791–94* (Urbana: University of Illinois Press, 1984). Of the three, combat motivation has proved the most difficult to study. Source material describing the experience of battle itself is more scarce: especially in conflicts before the twentieth century, accounts of combat are far less common than depictions of camp and campaign life (and since many soldiers intended their letters to reassure loved ones about their prospects, the resulting writings often avoided unnecessarily harrowing descriptions of battle). Too, there is the writer's separation from the events under examination: the bored soldier often chose to while away the hours in camp musing about the reasons for staying with the army, but every participant's description of the sensations of being in battle is necessarily removed (by minutes or hours, at best, and often by weeks, years, or decades) from the experience itself. The extreme conditions of the battlefield (sensory overload, confusion, and the presence of danger, suffering, and death) prove extraordinarily difficult to put into words. The powerful neurochemicals released into participants' systems during combat further skew both their perceptions and their memories. And the supremely lethal violence of the battlefield silences many of the voices researchers would like to hear most.

4. J. F. C. Fuller, *A Military History of the Western World*, vol. 2 (New York: Funk & Wagnalls, 1955), 196.

5. Elton E. Mackin, *Suddenly We Didn't Want to Die: Memoirs of a World War I Marine* (Novato, CA: Presidio Press, 1993), 29.

6. Europeans themselves noted national differences in the ways armies applied coercive motivators to compel soldiers to fight. From the British perspective,

Prussians trained men like "spaniels," with the cane; French armies relied instead on honor; and the British struck a happy medium, treating soldiers as thinking men. Sylvia R. Frey, *The British Soldier in North America: A Social History of Military Life in the Revolutionary Period* (Austin: University of Texas Press, 1981), 110–111. Motivating troops solely through physical coercion has often proved impossible in American armies. Frederick William Steuben, the Prussian officer who drilled the Continental Army into a disciplined force during the winter encampment of 1778, recognized that simply brutalizing American soldiers into submission (the common practice in contemporary European armies) was unlikely to succeed among the more independent-minded colonials and that other means of motivating the troops were therefore necessary. Accordingly, his drill manual asserted that the first object of a junior officer should be "to gain the love of his men, by treating them with every possible kindness and humanity," by "enquiring into their complaints, and when well founded, seeing them redressed." The junior officer, Steuben held, "should know every man of his company by name and character" and should procure for them "such comforts and conveniences as are in his power"—a far cry from the relationship European officers were advised to cultivate with their men. Frederick William Steuben, *Regulations for the Order and Discipline of the Troops of the United States* (Philadelphia: Styner & Cist, 1779), 135, 138. Fred Anderson's study of colonial troops' interactions with British regulars during the Seven Years' War argues that this cultural tension over the legitimacy of coercion predates the American revolution. Fred Anderson, *A People's Army: Massachusetts Soldiers and Society in the Seven Years' War* (Chapel Hill: University of North Carolina Press, 1984), 111–141.

7. Mary Livermore, *My Story of the War: A Woman's Experience* (Hartford, CT: A. D. Worthington, 1890), 559.

8. S. L. A. Marshall, *Men against Fire: The Problem of Battle Command in Future War* (New York: William Morrow, 1947); Edmund Shils and Morris Janowitz, "Cohesion and Disintegration in the Wehrmacht in World War II," *Public Opinion Quarterly* 12, no. 2 (Summer 1948): 280–315; Samuel Stouffer et al., *The American Soldier* (Princeton, NJ: Princeton University Press, 1949).

9. Marshall, *Men against Fire*, 42.

10. Nora Kinzer Stewart, *Mates and Muchachos: Unit Cohesion in the Falklands/Malvinas War* (New York: Brassey's [U.S.], 1991).

11. James M. McPherson, *For Cause and Comrades: Why Men Fought in the Civil War* (New York: Oxford University Press, 1997).

12. William Darryl Henderson, *Why the Vietcong Fought: A Study of Motivation and Control in a Modern Army in Combat* (Westport, CT: Greenwood Press, 1979), xv–xx.

13. Stephen E. Ambrose, *Citizen Soldiers: The U.S. Army from the Normandy Beaches to the Bulge to the Surrender of Germany, June 7, 1944–May 7, 1945* (New York: Simon & Schuster, 1997), 14.

14. Leonard Wong et al., *Why They Fight: Combat Motivation in the Iraq War* (Carlisle Barracks, PA: Strategic Studies Institute, U.S. Army War College, 2003), 23–25.

15. Wong et al., *Why They Fight*, 10, 11, 12.

16. Richard H. Kohn, "The Social History of the American Soldier: A Review and Prospectus for Research," *American Historical Review* 86, no. 3 (June 1981): 562.

17. Of course, not every soldier fought. Each of the wars examined in these pages involved soldiers whose will to fight wavered or broke at some point. American units broke in several battles during the War of Independence (at Brooklyn, Kip's Bay, and White Plains, to name a few), and one mutinied in Pennsylvania in 1781. Enough soldiers refused to fight in the mid-nineteenth century that Civil War troops created a panoply of colorful nicknames for those who avoided battle: skulkers, shirkers, and coffee-coolers, among others. Nor did every soldier in the Second World War muster the will to fight: in one well-known example, two full regiments of the 106th Infantry Division surrendered en masse during the Battle of the Bulge in December 1944. Robert Middlekauff, "Why Men Fought in the American Revolution," *Huntington Library Quarterly* 43 (Spring 1980): 135; Don Higginbotham, *The War of American Independence: Military Attitudes, Policies, and Practice, 1763–1789* (New York: Macmillan, 1971), 403–405; McPherson, *For Cause and Comrades*, 6–8; Gerhard Weinberg, *A World at Arms: A Global History of World War II* (Cambridge: Cambridge University Press, 1994), 767–769.

18. Russell Weigley, *History of the United States Army* (New York: Macmillan, 1967), 438. Even if each of these regiments initially went into combat as a true "band of brothers" forged through common experience in training, the units could not have sustained their will to fight over months if the emotional bonds between soldiers who knew each other well were the only thing motivating them in battle.

19. For example, Gerhard Weinberg has estimated that the German Army executed tens of thousands of its own troops during the Second World War for their reluctance to fight; the United States Army, in contrast, executed a single soldier, Eddie Slovik, for desertion.

20. Ernst Jünger, *Storm of Steel* (London: Chatto and Windus, 1929).

21. Ivan Bloch, *Is War Now Impossible? Being an Abridgment of "The War of the Future in Its Technological, Economic, and Political Relations"* (London: Grant Richards, 1899).

22. Stephen Biddle, *Military Power: Explaining Victory and Defeat in Modern Battle* (Princeton, NJ: Princeton University Press, 2004), 78–107.

23. Charles Ardant du Picq, *Battle Studies: Ancient and Modern War* (Harrisburg, PA: Military Service Publishing, 1947), 116.

24. William Slim, *Defeat into Victory* (New York: David McKay, 1961), 451.

25. S. L. A. Marshall, *Battle at Best* (New York: William Morrow, 1963), 272.

26. David Perry, "Recollections of an Old Soldier," *Magazine of History* 137 (1928): 9.

27. John Keegan, *The Face of Battle* (New York: Viking, 1976), 114–116, 274–284.

28. Anthony Kellett, *Combat Motivation: The Behavior of Soldiers in Battle* (Boston: Kluwer-Nijhoff, 1982), 334.

29. Roscoe C. Blunt, Jr., *Foot Soldier: A Combat Infantryman's War in Europe* (Cambridge, MA: Da Capo Press, 2001), 181.

30. John B. Babcock, *Taught to Kill: An American Boy's War from the Ardennes to Berlin* (Washington, DC: Potomac Books, 2005), 54.

31. Stouffer, *American Soldier*, 2:88. The transference on display is interesting: to some soldiers, it was their own army, rather than the enemy, that appeared the most proximate threat to survival.

32. Lewis Mumford, *Green Memories: The Story of Geddes Mumford* (New York: Harcourt, Brace, 1947), 306–307.

33. Hanson W. Baldwin, "The Tunisian Campaign," *Life*, June 14, 1943, 86.

34. Lewis M. Hosea, "The Second Day at Shiloh," in *Sketches of War History, 1861–1865*, vol. 6 (Cincinnati: R. Clark, 1908), 199–200.

35. Alpheus S. Williams, *From the Cannon's Mouth*, ed. Milo P. Quaife (Detroit: Wayne State University Press, 1959), 196.

36. Babcock, *Taught to Kill*, 81.

37. Samuel Hynes, *The Soldiers' Tale: Bearing Witness to Modern War* (New York: Viking, 1997), 8.

Chapter 1. The Evolving Character of Infantry Combat

1. Roscoe C. Blunt, Jr., *Foot Soldier: A Combat Infantryman's War in Europe* (Cambridge, MA: Da Capo Press, 2001), 183.

2. Ross S. Carter, *Those Devils in Baggy Pants* (New York: Appleton-Century-Crofts, 1951), 138.

3. David Tucker Brown, Jr., to mother, 18 December 1942, in David Tucker Brown, Jr., *Letters of a Combat Marine* (Chapel Hill: University of North Carolina Press, 1947), 20.

4. Leon C. Standifer, *Not in Vain: A Rifleman Remembers World War II* (Baton Rouge: Louisiana State University Press, 1992), 101.

5. Edward C. Arn, *Arn's War: Memoirs of a World War II Infantryman, 1940–1946*, ed. Jerome Mushkat (Akron: University of Akron Press, 2006), 86.

6. Oliver Wendell Holmes, Jr., "The Soldier's Faith," address delivered 30 May 1895, at Harvard University, in *The Essential Holmes: Selections from the Letters, Speeches, Judicial Opinions, and Other Writings of Oliver Wendell Holmes, Jr.*, ed. Richard A. Posner (Chicago: University of Chicago Press), 87–93.

7. Anthony Swofford, *Jarhead: A Marine's Chronicle of the Gulf War and Other Battles* (New York: Scribner, 2003), 5–7. The dissonance of that image—Marines looking to Vietnam-era films for inspiration and insight on the eve of a war that Americans at home were being reassured would not devolve into precisely that kind of quagmire—was not lost on Swofford and his comrades.

8. John Keegan, *The Face of Battle* (New York: Viking, 1976), 303.

9. Charles Ardant du Picq, *Battle Studies: Ancient and Modern War* (Harrisburg, PA: Military Service Publishing, 1947), 109.

10. Field Manual 100-5, *Field Service Regulations* (Washington, DC: U.S. Government Printing Office, 1941), 18.

11. Samuel Webb to Joseph Webb, 19 June 1775, in *A Salute to Courage: The*

American Revolution as Seen through Wartime Writings of Officers of the Continental Army and Navy, ed. Dennis P. Ryan (New York: Columbia University Press, 1979), 6.

12. John T. McMahon diary entry, 15 May 1864, in John T. McMahon, *John T. McMahon's Diary of the 136th New York, 1861–1864*, ed. John Michael Priest (Shippensburg, PA: White Mane, 1993), 94.

13. Donald R. Burgett, *Seven Roads to Hell: A Screaming Eagle at Bastogne* (Novato, CA: Presidio Press, 1999), 12.

14. Alvin Boeger to brother John Boeger, 20 November 1944, 99th Division Collection, U.S. Army Military History Institute, Carlisle, Pennsylvania (additional archival sources from this collection will be referred to as MHI).

15. Jonathan Brigham, 1832 pension application; William Hutchinson, 1836 pension application, in John C. Dann, ed., *The Revolution Remembered: Eyewitness Accounts of the War for Independence* (Chicago: University of Chicago Press, 1980), 3–4, 146.

16. Reuben Kelly to sister, 15 May 1863, Kelly Letters, Wisconsin Historical Society, Madison, Wisconsin (hereafter cited as WHS). Many soldiers quoted in these pages, particularly those from the eighteenth and nineteenth centuries, employed idiosyncratic spelling and grammar in their writing. Rather than regularize these constructions, this work retains the spelling and syntax found in the original documents.

17. Colonel J. J. Scales in Grady McWhiney and Perry Jamieson, *Attack and Die: Civil War Military Tactics and the Southern Heritage* (Tuscaloosa: University of Alabama Press, 1982), 11.

18. Harry Bare in Ronald J. Drez, ed., *Voices of D-Day: The Story of the Allied Invasion Told by Those Who Were There* (Baton Rouge: Louisiana State University Press, 1994), 207.

19. Harry S. Arnold, "'Easy' Memories: The Way It Was," unpublished manuscript, 99th Infantry Division Collection, MHI, 31.

20. As one Union soldier wrote, "We hear a great deal about hand-to-hand fighting. Gallant though it would be, and extremely pleasant to the sensation newspapers to have it to record, it is of very rare occurrence," since "when men can kill one another at six hundred yards they generally would prefer to do it at that distance than to come down to two paces." Henry Otis Dwight, "How We Fight at Atlanta," *Harper's New Monthly Magazine* 29 (October 1864): 665. A half-century later, an American Marine endorsed that revulsion at up-close combat: "The prospect of using a bayonet, of facing enemy bayonets in action, was his pet horror. The very thought left him weak." Elton E. Mackin, *Suddenly We Didn't Want to Die: Memoirs of a World War I Marine* (Novato, CA: Presidio Press, 1993), 28. David Grossman's *On Killing* takes this natural resistance to killing as its core theme, examining the ways that the U.S. Army engineered its training in the twentieth century to help overcome soldiers' innate aversion to the act. David Grossman, *On Killing: The Psychological Cost of Learning to Kill in War and Society* (New York: Little, Brown, 1995).

21. The confusion of battle and the impossibility of any one participant understanding all of it from within also facilitated the creation and spread of rumors,

another continuity that persisted from war to war. Such rumors ran the gamut from the hopeful (the day is won; fresh replacements are on the way; a critical enemy leader has been wounded, killed, or captured) to the calamitous (the line is flanked; enemy soldiers have infiltrated the rear). A Wisconsin veteran who spent the third day of the battle of Gettysburg laying with his regiment in their breastworks far away from the main action of the day's fighting nonetheless confidently (and erroneously) reported in his diary that "[Confederate General James] Longstreet is a prisoner." George W. Downing journal entry, 3 July 1863, WHS. Paul Fussell devotes a chapter, "Rumors of War," to this phenomenon in his book *Wartime*, citing a variety of rumors (saltpeter is being added to rations to render troops more docile; the first hundred Marines to set foot on the Japanese home islands will receive a free Ford car), which Fussell attributed in part to the need to impose a sense of meaning and to assert some control over the experience of being at war. Paul Fussell, *Wartime: Understanding and Behavior in the Second World War* (New York: Oxford University Press, 1989), 35–52.

22. Horace Smith diary entry, 1 May 1863, WHS.

23. Anonymous Confederate soldier to wife, 17 April 1862, in Bell I. Wiley, *The Life of Johnny Reb, Common Soldier of the Confederacy* (Baton Rouge: Louisiana State University Press, 1943), 34.

24. David Smith to wife, 8 August 1863, in *The Soldier's Pen: Firsthand Impressions of the Civil War*, ed. Robert E. Bonner (New York: Hill and Wang, 2006), 79.

25. Ernie Pyle, *Here Is Your War* (New York: Henry Holt, 1943), 151.

26. Don Higginbotham briefly discusses the battle and its context within the overall conflict in *The War of American Independence: Military Attitudes, Policies, and Practice, 1763–1789* (New York: Macmillan, 1971), 366–367. Lawrence Babits furnishes a much more exhaustive description of the fight itself in *A Devil of a Whipping: The Battle of Cowpens* (Chapel Hill: University of North Carolina Press, 1998).

27. Babits, *Devil of a Whipping*, 11–13.

28. Ibid., 104.

29. Thomas Young, "Memoir of Major Thomas Young," *Orion* 3 (1843): 100.

30. Ibid.

31. Babits, *Devil of a Whipping*, 160.

32. James McPherson provides a brief synopsis of the context and significance of the Shiloh campaign in *Battle Cry of Freedom: The Civil War Era* (New York: Oxford University Press, 1988), 405–416. James Lee McDonough treats the campaign and the battle in much more detail in *Shiloh: In Hell before Night* (Knoxville: University of Tennessee Press, 1977). Particularly useful for its bottom-up examination of soldiers' experiences in combat at Shiloh is Joseph Allan Frank and George A. Reaves's *"Seeing the Elephant": Raw Recruits at the Battle of Shiloh* (Champaign: University of Illinois Press, 2003), which places heavy emphasis on the perspectives and memories of the untested troops who contested the two-day battle.

33. Leander Stillwell, *The Story of a Common Soldier of Army Life in the Civil War, 1861–1865* (Kansas City, MO: Franklin Hudson, 1920), 38.

34. Charles Royster, *A Revolutionary People at War: The Continental Army and American Character, 1775–1783* (Chapel Hill: University of North Carolina Press, 1979), 132.

35. In his annual report for 1855, Secretary of War Jefferson Davis indicated that Federal arsenals had already ceased production of smoothbores entirely and would begin the process of converting production to rifled muskets by year's end. McWhiney and Jamieson, *Attack and Die*, 48–49.

36. Charles T. Morey to sister, 12 January 1863, Charles Morey Letters, Stuart Goldman Collection, MHI.

37. S. H. M. Byers, "How Men Feel in Battle: Recollections of a Private at Champion Hills," *Annals of Iowa* 2 (July 1896): 443.

38. Babits, *Devil of a Whipping*, 87–88.

39. Joshua K. Callaway to Dulcinea Callaway, 24 September 1863, in Joshua K. Callaway, *The Civil War Letters of Joshua K. Callaway*, ed. Judith Lee Hallock (Athens: University of Georgia Press, 1997), 137.

40. David E. Beem to fiancée, 14 December 1862, David E. Beem Papers, Indiana Historical Society (hereafter cited as IHS).

41. George Hurlbut diary entry, 8 April 1862, in Frank and Reaves, *Seeing the Elephant*, 91.

42. Augustus Van Dyke to wife, 21 September 1862, Augustus Van Dyke Papers, IHS.

43. John T. McMahon diary entry, 16 November 1863, in McMahon, *Diary*, 70.

44. Ulysses S. Grant, *Personal Memoirs of Ulysses S. Grant* (Cambridge, MA: Da Capo Press, 1982), 178. Indeed, as the Civil War wore on, the cavalry charge all but disappeared, and troopers generally fought dismounted as infantry.

45. Stillwell, *Common Soldier*, 54.

46. Grant, *Personal Memoirs*, 178. According to Grant, Sherman also had a series of horses shot out from beneath him. Grant argued that Sherman's "constant presence" among his troops "inspired a confidence in officers and men that enabled them to render services on that bloody battle-field worthy of the best of veterans."

47. McDonough, *Shiloh*, 152–153.

48. Ibid., 162–167; McPherson, *Battle Cry*, 410.

49. Grant, *Personal Memoirs*, 181.

50. Grant's medical director J. H. Brinton told of thousands of victims at Shiloh, "wounded and lacerated in every conceivable manner, on the ground, under a pelting rain, without shelter, without bedding, without straw to lay upon, and with very little food." The largest portion of the wounded had been shot in the extremities; men shot in the head and abdomen, Brinton observed, usually died of their injuries. The "agonies of the wounded," he noted, "were beyond all description." U.S. Army Surgeon General's Office, *The Medical and Surgical History of the War of the Rebellion, 1861–1865* (Washington, DC: U.S. Government Printing Office, 1870), Medical Volume, part 1, appendix, 31.

51. Lester Atwell, *Private* (New York: Simon & Schuster, 1958), 22.

52. Carter, *Devils*, 170–172.

53. Ernst Jünger, *Storm of Steel* (London: Chatto and Windus, 1929).

54. John Babcock, *Taught to Kill: An American Boy's War from the Ardennes to Berlin* (Washington, DC: Potomac Books, 2005), 63.

55. Arnold, "'Easy' Memories," 47.

56. Edward G. Miller supplies an excellent overview of the campaign in *A Dark and Bloody Ground: The Hürtgen Forest and the Roer River Dams, 1944–1945* (College Station: Texas A&M University Press, 1995). Robert S. Rush's *Hell in Hürtgen Forest: The Ordeal and Triumph of an American Infantry Regiment* (Lawrence: University Press of Kansas, 2001) provides a more detailed ground-level view of the fighting in the Hürtgen, recreating the battle experiences of the American 22nd Infantry Regiment and examining in detail the combat that cost that unit such severe casualties and the factors that enabled them to persevere despite their staggering losses.

57. One World War II GI recalled the tangible presence of the storm of steel from the perspective of the individual soldier hugging the ground: "You could have held your hand up and probably stopped a tracer bullet. You could see the tracers going about four feet over your head." John Zmudzinski in Drez, *Voices*, 239.

58. Peter R. Mansoor, *The GI Offensive in Europe: The Triumph of American Infantry Divisions, 1941–1945* (Lawrence: University Press of Kansas, 1999), 187.

59. Standifer, *Not in Vain*, 52. That emphasis on stealth, and the fear that any careless talk or action would draw unwanted attention, pervaded every facet of the World War II soldier's experience. Combat journalist Ernie Pyle noted that soldiers in Europe "couldn't rise even for nature's calls" since the enemy "felt for them continually with his artillery." Pyle, *Here Is Your War*, 151.

60. William Devitt, *Shavetail: The Odyssey of an Infantry Lieutenant in World War II* (St. Cloud, MN: North Star Press, 2001), 75.

61. Stephen Biddle, *Military Power: Explaining Victory and Defeat in Modern Battle* (Princeton, NJ: Princeton University Press, 2004), 31–33.

62. Ernest Hemingway, who observed the battle in the Hürtgen as a journalist, described it in terms of a World War I battle famous for its savagery: "Passchendale with tree bursts."

63. Burgett, *Seven Roads*, 155.

64. Paul Fussell in Richard M. Stannard, *Infantry: An Oral History of a World War II American Infantry Battalion* (New York: Twayne, 1993), 104.

65. Arn, *Arn's War*, 32.

66. Rush, *Hell in Hürtgen Forest*, 147, 215.

67. Pyle, *Here Is Your War*, 153. Robert Rush describes the agony of dealing with mines during the battle in Hürtgen Forest in vivid detail. The German defenders had sowed nearly every road and trail with mines, making them all but impassable; many of the mines were stacked two and three deep, and mines buried atop one another frustrated engineers' attempts to remove them. Even after

multiple sweeps some of the mines remained deep in the mud, only to explode as American vehicles attempted to make their way through. Rush, *Hell in Hürtgen Forest*, 162–163, 192. Later chapters in this book explore the costs of maintaining such constant vigilance in more detail.

68. Donald Faulkner in Rush, *Hell in Hürtgen Forest*, 222.

69. Peter Kindsvatter discusses the new category of twentieth-century wounds (and their British antecedent from the First World War, the "blighty") in *American Soldiers: Ground Combat in the World Wars, Korea, and Vietnam* (Lawrence: University Press of Kansas, 2003), 121–123.

70. Gerald Nelson in Gerald Astor, *A Blood-Dimmed Tide: The Battle of the Bulge by the Men Who Fought It* (New York: Donald J. Fine, 1992), 161.

71. Donald Faulkner in Rush, *Hell in Hürtgen Forest*, 222.

72. Babcock, *Taught to Kill*, 36.

73. A British NCO grumbling about artillery in the Falklands War complained that he could handle a sniper—another person within the range of his own weapon—firing at him, but "what do you do about some fucker four miles away?" Quoted in Richard Holmes, *Acts of War: The Behavior of Men in Battle* (New York: Free Press, 1985), 211.

74. S. L. A. Marshall, *Battle at Best* (New York: William Morrow, 1963), 272. Samuel Stouffer's interviews with World War II veterans suggested that that feeling of isolation was emotional as well as literal: "Each man had the inner loneliness that comes from having to face death at each moment." Samuel Stouffer, *The American Soldier*, vol. 2, *Combat and Its Aftermath* (Princeton, NJ: Princeton University Press, 1949), 99.

75. Burgett, *Seven Roads to Hell*, 76–77.

76. Arnold, "'Easy' Memories," 47.

77. Ardant du Picq, *Battle Studies*, 116. The next chapter explores the effects of that isolation on soldiers' ability to manage their fears and the effects of isolation amid the terror of battle.

78. Byers, "How Men Feel in Battle," 443.

79. Earl Hess's analysis of Union soldiers' experiences in mid-nineteenth-century battle argued that for the individual in the firing line, getting hit appeared largely a matter of chance; near misses, according to Hess, were far more common than lethal strikes. Earl J. Hess, *The Union Soldier in Battle: Enduring the Ordeal of Combat* (Lawrence: University Press of Kansas, 1997), 24. That was the experience of Civil War volunteer Henry Welch. Standing behind a rail fence, he had just spotted a pair of enemy soldiers behind a bush when a ball struck him. It was not a lethal hit; the ball thumped his finger, tearing off the nail, then bounced off his arm and passed out his sleeve. Writing his aunt and uncle, Welch surmised that the ball would have done far more damage if it had hit him "fair." Henry Welch to Polly and Franklin Tanner, 10 May 1863, Henry Welch Papers, MHI.

80. Hillory Shifflet to wife, 26 October 1862, in Bonner, *The Soldier's Pen*, 93.

81. Jack Keating in Drez, *Voices*, 276.

82. Robert Crisp, *The Gods Were Neutral* (London: Frederick Muller, 1960), 12.

83. William Dunfee in Astor, *Blood-Dimmed Tide*, 377.

84. Carter, *Devils*, 174.

85. Theodore Aufort in Drez, *Voices*, 241.

86. Ernie Pyle remarked on the nocturnal activities of soldiers constantly threatened with the possibility of enemy attack: "Each outfit was provided with the password beforehand. In the shadows, soldiers couldn't tell who was who, and everyone was afraid of getting shot by his own men, so all night the hillsides around Oran hissed with the constantly whispered password directed at every approaching shadow." Pyle, *Here Is Your War*, 23.

87. A. W. Stillwell diary entry, 2 July 1863, WHS.

88. Henry Welch to Polly and Franklin Tanner, 9 June 1864, Henry Welch Papers, MHI.

89. Ebeneezer Hannaford, "In the Ranks at Stones' River," in *Battles and Leaders of the Civil War*, ed. Peter Cozzens, vol. 6 (Champaign: University of Illinois Press, 2006), 182.

90. Chet Cunningham, *Hell Wouldn't Stop: An Oral History of the Battle of Wake Island* (New York: Carroll & Graf, 2002), 22–23.

91. Astor, *Blood-Dimmed Tide*, 138.

92. Bill Davidson, *Cut Off: Behind Enemy Lines in the Battle of the Bulge with Two Small Children, Ernest Hemingway, and Other Assorted Misanthropes* (New York: Stein and Day, 1972), 25–26.

93. Ardant du Picq, *Battle Studies*, 115.

94. Blunt, *Foot Soldier*, 1. Peering at the graves of soldiers in Europe, scattered with wildflowers picked from nearby fields, Blunt ruminated on his own mortality: "Was this to be my future?" Ibid., 94.

95. William S. Powell, "A Connecticut Soldier under Washington: Elisha Bostwick's Memoirs of the First Years of the Revolution," *William and Mary Quarterly*, 3rd ser., 6 (1949): 101.

96. David Hunter Strother diary entries, 24 March 1862 and 25 March 1862, in Strother, *Virginia Yankee*, 19–20.

97. Blunt, *Foot Soldier*, 94.

98. As Martin van Creveld put it in his comparison of German and American army effectiveness during the Second World War, it is "a matter of definition" that "no kind of utilitarian reasoning in the world can render the individual willing to lay down HIS life," even though such logic might provide an eminently sound basis for a nation's decision to go to war. Martin van Creveld, *Fighting Power: German and U.S. Army Performance, 1939–1945* (Westport, CT: Greenwood Press, 1982), 170.

99. Alfred Bell to wife, 11 January 1863, 19 August 1864, Bell Papers, Perkins Library, Duke University, Durham, North Carolina.

100. Stillwell, *Common Soldier*, 44.

101. A. N. Erskine to wife, 28 June 1862, in Wiley, *Johnny Reb*, 32.

102. Charles Wellington Reed to mother, 13 August 1863, in Charles Wellington Reed, *"A Grand Terrible Dramma" from Gettysburg to Petersburg: The Civil War*

Letters of Charles Wellingon Reed, ed. Eric A. Campbell (New York: Fordham University Press, 2000), 128.

Chapter 2. Fear in Combat

1. Harry Bare in Ronald J. Drez, ed., *Voices of D-Day: The Story of the Allied Invasion Told by Those Who Were There* (Baton Rouge: Louisiana State University Press, 1994), 207.

2. Alvin Boeger Questionnaire, 99th Division Collection, U.S. Army Military History Institute, Carlisle, Pennsylvania (hereafter referred to as MHI).

3. Samuel Webb to brother Joseph Webb, 19 June 1775, in Dennis P. Ryan, ed., *A Salute to Courage: The American Revolution as Seen through the Wartime Writings of Officers of the Continental Army and Navy* (New York: Columbia University Press, 1979), 6.

4. Harold J. Gordon, Jr., *One Man's War: A Memoir of World War II,* ed. Nancy M. Gordon (New York: Apex Press, 1999), 17, 25.

5. Leon C. Standifer, *Not in Vain: A Rifleman Remembers World War II* (Baton Rouge: Louisiana State University Press, 1992), 145.

6. Eugene Sledge, *With the Old Breed at Peleliu and Okinawa* (Novato, CA: Presidio Press, 1981), 69.

7. Studies of pilots in the Eighth Air Force during the Second World War, for example, found that on early missions, aircrew often feared not functioning well on the mission or being thought a coward; on later missions, fears of wounding and death replaced those fears of personal failure. S. J. Rachman, *Fear and Courage* (San Francisco: W. H. Freeman, 1978), 52.

8. In *For Cause and Comrades,* James McPherson quotes soldiers who asserted that they never experienced fear in battle. The author quotes one Virginia cavalryman who asserted in a letter "I never felt fear at all": the trooper was "getting so used to cannon balls flying over" his head that he no longer minded them. McPherson attributes some of those statements to bravado or denial; in the main, soldiers who experienced the danger of shots fired in anger generally confessed to feelings of fear at one point or another. James McPherson, *For Cause and Comrades: Why Men Fought in the Civil War* (New York: Oxford University Press, 1997), 36–38.

9. Abner Small, *The Road to Richmond: The Civil War Memoirs of Major Abner R. Small,* ed. Harold A. Small (Berkeley: University of California Press, 1959), 70.

10. Jerry Nelson and Joe Coomer in Gerald Astor, *A Blood-Dimmed Tide: The Battle of the Bulge by the Men Who Fought It* (New York: Donald I. Fine, 1992), 161, 204.

11. Samuel Stouffer et al., *The American Soldier,* vol. 2, *Combat and Its Aftermath* (Princeton, NJ: Princeton University Press, 1949), 200–202. William Manchester relates a similar phenomenon in a possibly apocryphal episode from his memoir *Goodbye, Darkness.* William Manchester, *Goodbye, Darkness: A Memoir of the Pacific War* (Boston: Little, Brown, 1979), 3–7.

12. Ernie Pyle, *Here Is Your War* (New York: Henry Holt, 1943), 23. Other soldiers suffered physical symptoms of a different nature: one twenty-five-year-old Civil War captain experienced a more dramatic manifestation of fear, writing his mother that his hair had gone "quite gray," but admitting that the change was minor given "the hardships of the past summer." Charles C. Morey to mother, 27 January 1865, Morey Letters, MHI.

13. Rice C. Bull, *Soldiering: The Civil War Diary of Rice C. Bull, 123rd New York Volunteer Infantry,* ed. Jack K. Bauer (San Rafael, CA: Presidio Press, 1977), 41.

14. Eugene Schroeder in Drez, *Voices,* 187.

15. George Rowland to brother, 26 November 1864, Rowland File, MHI.

16. Harry S. Arnold, "'Easy' Memories: The Way It Was," unpublished manuscript, 99th Division Collection, MHI, 31.

17. Bill Mauldin, *Bill Mauldin's Army* (Novato, CA: Presidio Press, 1979), 163.

18. Bull, *Soldiering,* 55.

19. Ernie Pyle, *Last Chapter* (New York: Henry Holt, 1946), 99.

20. Gordon, *One Man's War,* 24.

21. Boeger Questionnaire.

22. Arnold, "'Easy' Memories," 74–75.

23. David Hunter Strother diary entry, 21 July 1864, in David Hunter Strother, *A Virginia Yankee in the Civil War: The Diaries of David Hunter Strother,* ed. Cecil D. Eby, Jr. (Chapel Hill: University of North Carolina Press, 1961), 281.

24. John Zmudzinski in Drez, *Voices,* 239.

25. Henry Welch to Polly and Franklin Tanner, 15 July 1863, Welch Papers, MHI.

26. Bruce Bradley in Drez, *Voices,* 183.

27. A. B. Isham, "The Story of a Gunshot Wound," in *Sketches of War History, 1861–1865,* vol. 4 (Cincinnati: R. Clark, 1896), 430. The sensations were so muted, in fact, that Isham initially could not determine whether he had been struck by one bullet or two.

28. Abel Potter, 1832 pension application, in John C. Dann, ed., *The Revolution Remembered: Eyewitness Accounts of the War for Independence* (Chicago: University of Chicago Press, 1980), 25.

29. Bull, *Soldiering,* 57.

30. Joshua K. Callaway to wife, 2 June 1862, in Joshua K. Callaway, *The Civil War Letters of Joshua K. Callaway,* ed. Judith Lee Hallock (Athens: University of Georgia Press, 1997), 24.

31. Bull, *Soldiering,* 118.

32. John Dollard, one of the first academics to study fear in battle, held that its chief value was the caution it instilled in troops under fire. John Dollard, *Fear in Battle* (New Haven, CT: Institute of Human Relations, 1943), 12.

33. Gordon, *One Man's War,* 19.

34. Harry Reisenleiter in Drez, *Voices,* 138. Distractions born of fear also led to potentially ruinous oversights: Leonard Griffing, a soldier who jumped into France on the eve of D-Day, realized at first light that the magazine in his carbine was missing. "In all the excitement," he found, "I'd thrown back the bolt to

chamber the first round and forgot to put in a magazine first." The confusion induced by fear had led him to spend the first night of the invasion with an empty weapon. Griffing linked his error to those made by earlier American soldiers in similar states of agitation: "I'm told that during the 1860 war, they found rifles after the battles that had as many as ten charges shoved into them, but the guy with the rifle had forgotten to pull the trigger even once." Leonard Griffing in Drez, *Voices*, 78.

35. Gordon, *One Man's War*, 18.

36. Dollard's veterans named hunger and thirst, exposure to excessive heat and cold, and prolonged fatigue as the critical stresses that rendered them less capable of fighting. Indeed, 83 percent of Dollard's veterans volunteered that "being tired made them poorer soldiers." Dollard, *Fear in Battle*, 51–52.

37. George Tillotson to wife, 22 May 1862, in Robert E. Bonner, ed., *The Soldier's Pen: Firsthand Impressions of the Civil War* (New York, NY: Hill and Wang, 2006), 93.

38. Frederick Heuring diary entry, 22 December 1862, Heuring Journal, Indiana Historical Society, Indianapolis, Indiana (hereafter referred to as IHS).

39. Gordon, *One Man's War*, 30.

40. The degree of deprivation varied according to the quality of the armies' logistical system. In one extreme example, Richard Vining, a soldier in the War of Independence, found himself forced to slaughter and eat a dog on one occasion; later he and his ravenous messmates killed and ate an owl. Richard Vining, 1833 pension application, in Dann, *Revolution Remembered*, 17.

41. Joseph Allan Frank and George Reaves, *"Seeing the Elephant": Raw Recruits at the Battle of Shiloh* (New York: Greenwood Press, 1989), 102. That overwhelming thirst is one of the reasons that so many Civil War photographs depicted soldiers half submerged in mud puddles: desperately thirsty dying men crawled to any source of water available to them as they expired.

42. S. H. M. Byers, *With Fire and Sword* (New York: Neale Publishing, 1911), 79.

43. Those natural counter-regulatory measures, centered on the hypothalamus gland, offer another example of the ways in which natural reactions to danger and fear echo one another, whether in the eighteenth century or the twentieth. The hypothalamus, after all, senses only that the body is losing blood volume; the gland cannot determine whether the reasons for that loss lay with solid shot from a British two-pound cannon, a Minié ball, or shrapnel from a German 88-millimeter gun—or, for that matter, a car accident.

44. Author Robert Rush, himself a veteran NCO, argued that, in his personal experience, "nothing is more enervating than to wake up day after day cold and wet. Uncontrollable shivers flow through the body and signal the onset of hypothermia." Robert S. Rush, *Hell in Hürtgen Forest: The Ordeal and Triumph of an American Infantry Regiment* (Lawrence: University Press of Kansas, 2001), 155.

45. Michael D. Doubler, *Closing with the Enemy: How GIs Fought the War in Europe, 1944–1945* (Lawrence: University Press of Kansas, 1994), 186–187.

46. Charles Wilson, *The Anatomy of Courage: The Classic World War I Study of the Psychological Effects of War* (London: Constable, 1945).

47. John B. Babcock, *Taught to Kill: An American Boy's War from the Ardennes to Berlin* (Washington, DC: Potomac Books, 2005), 119.

48. Gerald Linderman, *Embattled Courage: The Experience of Combat in the Civil War* (New York: Free Press, 1987); McPherson, *Cause and Comrades*, 77–84.

49. John Baynes, *Morale: A Study of Men and Courage* (New York: Frederick A. Praeger, 1967).

50. Dollard, *Fear in Battle*. Peter Kindsvatter discusses the fears of twentieth-century American combat soldiers from World War I to the Vietnam War in *American Soldiers: Ground Combat in the World Wars, Korea, and Vietnam* (Lawrence: University Press of Kansas, 2003), 49–50, 82–83.

51. Arnold, "'Easy' Memories," 31.

52. Bull, *Soldiering*, 56.

53. Dollard's study of Spanish Civil War veterans reached a similar conclusion: while fear was a powerful natural response to danger, it did not necessarily dictate behavior. Dollard, *Fear in Battle*, 12.

54. Garrett Watts, 1834 pension application, in Dann, *Revolution Remembered*, 195.

55. Donald Burgett, *Seven Roads to Hell: A Screaming Eagle at Bastogne* (Novato, CA: Presidio Press, 1999), 42.

56. Arnold, "'Easy' Memories," 31.

57. Charles W. Bardeen, *A Little Fifer's War Diary* (Syracuse, NY: Printed by the author, 1910), 106.

58. Bull, *Soldiering*, 50. Though dispersed tactics increased the amount of space between friendly soldiers and blunted the power of directly observed behavior, even twentieth-century GIs witnessed the power of courage acquired by example. One Marine who fought at Peleliu—crossing a field pocked with artillery and mortar bursts and machine-gun fire—encountered a comrade years later at a reunion who told him, "I was scared to death, but I saw you running across, and I thought if you could do it, I could do it." Peter Richmond, *My Father's War: A Son's Journey* (New York: Simon & Schuster, 1996), 213.

59. John Dooley, *John Dooley, Confederate Soldier: His War Journal*, ed. Joseph T. Durkin (Washington, DC: Georgetown University Press, 1945), 46–47.

60. Bryan Grimes to wife, 20 September 1864, Grimes Papers, Southern Historical Collection, University of North Carolina–Chapel Hill. Though McPherson downplays the importance of coercion in driving the soldiers of the Civil War (his title, *For Cause and Comrades*, tellingly overlooks compulsion in its formulation of those troops' motivations in battle), the book nonetheless collects a number of memorable instances in which the stick wielded considerably more influence than the carrot. One came from an Indiana soldier who overheard General William T. Sherman reprimand a regiment for its cowardly performance for an hour, before "promising them that at the next battle they should be put in the foremost rank with a battery of Artillery immediately behind them and then if they attempted to run they would open on them with grape and canister." McPherson, *Cause and Comrades*, 48–52.

61. Garrett Watts, 1834 pension application, in Dann, *Revolution Remembered*, 195.

62. E. L. Marsh, "Military Discipline," in *War Sketches and Incidents*, vol. 2 (Des Moines, IA: n.p., 1898), 107; quoted in Earl J. Hess, *The Union Soldier in Battle: Enduring the Ordeal of Combat* (Lawrence: University Press of Kansas, 1997), 110. Hess's detailed study of Union soldiers in battle suggests that this intimate sensation of contact had perhaps the single greatest beneficial effect upon infantrymen's morale.

63. William Thompson Lusk, *War Letters of William Thompson Lusk, Captain, Assistant Adjutant-General, United States Volunteers 1861–1863* (New York: Private printing, 1911), 247–248.

64. John William DeForest, *A Volunteer's Adventures: A Union Captain's Record of the Civil War*, ed. James Croushore (New Haven, CT: Yale University Press, 1946), 60.

65. Burgett, *Seven Roads*, 61

66. Charles Ardant du Picq, *Battle Studies: Ancient and Modern War* (Harrisburg, PA: Military Service Publishing, 1947), 116.

67. Ibid., 114.

68. Richard Holmes, *Acts of War: The Behavior of Men in Battle* (New York: Free Press, 1985), 140.

69. Seventy percent of the veterans in Dollard's sample found it comforting to know that others in the unit were afraid, also: such recognition blunted concern of being judged negatively and increased the sense of cohesion within the group. Eighty-five percent of these veterans favored open discussion of fear within a unit preparing for battle. Dollard, *Fear in Battle*, 38–39.

70. Ardant du Picq, *Battle Studies*, 115.

71. S. L. A. Marshall, *Men against Fire: The Problem of Battle Command in Future War* (New York: William Morrow, 1947), 22.

72. Participants in one experiment, for example, were wired with sensors that measured physiological indicators of fear (heart rate, skin conductivity, and respiration) before embarking on their first parachute jump. Participants described their perceived feelings of fear during the exercise as the sensors recorded their physical responses. Their subjective impressions of their own levels of fear at various points in the exercise had little correlation with their biological symptoms. The objective measures indicated a slowly, steadily increasing level of tension from the arrival at the airport to takeoff to the jump, easing off gradually after the chute opened. Self-reported levels of fear, on the other hand, spiked upon arrival at the airport, dipped dramatically once the plane left the runway, and rose sharply again after exiting the plane, peaking just before the chute opened. Rachman, *Fear and Courage*, 22.

73. Ibid., 51.

74. Martin Seligman, *Learned Helplessness: A Theory for the Age of Personal Control* (New York: Oxford University Press, 1993). As with fear, the data suggest that it is the perception of controllability, more than the objective reality of control,

that produced the beneficial sense of self-efficacy. Studies of aircrew in the Second World War reinforced the validity of these observations on fliers in combat. In bomber crews, for example, the rate of psychological breakdown due to combat stress was much higher among gunners, who exerted no control over the plane's course or its evasive maneuvers, than among pilots, who exercised at least some control over the aircraft. The relative rates of psychological breakdown among bomber crews and fighter pilots suggested a similar pattern. Bomber crews suffered breakdown at a rate four times that of fighter crews: an unusual finding in the sense that, statistically, both groups' chances of survival were equally dismal during the period in question. The difference, it seemed, owed to the perception of control, or the lack of it: fighter pilots, whose sense of control over their nimble planes (and thus their chances in aerial dogfights) was far greater than that of bomber pilots (who complained frequently of being resigned to fly slow and level while enduring attacks by flak and enemy aircraft), enjoyed a substantial psychological benefit over their bomb-laden colleagues, as well. Rachman, *Fear and Courage*, 52. Intriguing experiments performed on canines suggest that humans are not alone in experiencing the demoralizing effects of helplessness when placed in situations over which they perceive no control. Seligman describes one such investigation in *Learned Helplessness*, in which experimenters placed two group of dogs in pens divided by a low wall, one side of which featured a floor rigged to trigger electrical shock. One group of dogs experienced a clear pattern of shocks: standing on one side of the pen activated the painful sensation, and the dogs quickly learned to jump over the wall to the get to the other side of the pen. Later, when the shocks were randomized, members of that first group continued to jump back and forth over the wall in search of relief, apparently conditioned to view their actions as providing some chance to end the misery. Experimenters placed a second group of dogs in an enclosure that delivered random shocks to both sides of the pen: jumping over the wall did not guarantee relief. That group quickly exhibited a kind of conditioned helplessness and gave up attempting to avoid the shocks altogether: when the jolts were later regularized to one side of the pen, researchers discovered that the demoralized dogs refused to jump over the wall even to test the other side, apparently convinced that their actions had little power to alleviate their suffering.

75. It is testimony to the power of those patterns on the individual understanding of the world that when such patterns are absent, subjects imagine them. The 2008 article "Lacking Control Increases Illusory Pattern Perception" marshaled more data to support that conclusion. The authors note that a "desire to combat uncertainty and maintain control has long been considered a primary and fundamental motivating force in human life and one of the most important variables governing psychological well-being and physical health." Subjects who can control (or perceive that they can control) the length of a painful shock during experiments register lower arousal, for example. Conversely, the authors note, perceiving a lack of control is an "unsettling and aversive state" that activates a fear response in the amygdala, which forms the most important reason that individuals

try to establish a sense of control even when it vanishes or is taken from them. The authors argue that when individuals are "unable to gain a sense of control objectively, they will try to gain it perceptually." That is, in situations that seem beyond individual control, "people will turn to pattern perception, the identification of a coherent and meaningful interrelationship among a set of stimuli." That pattern perception helps individuals make sense of events; lacking control leads people to identify a "coherent and meaningful relationship" even "among a set of random or unrelated stimuli." Jennifer A. Whitson and Adam D. Galinsky, "Lacking Control Increases Illusory Pattern Perception" *Science* 322, no. 115 (October 2008): 115–117.

76. Babcock, *Taught to Kill*, 60–61.

77. Sam R. Watkins, *"Co. Aytch": A Side Show of the Big Show* (New York: Macmillan, 1962), 29.

78. Garrett Watts, 1834 pension application, in Dann, *Revolution Remembered*, 194.

79. Dollard, *Fear in Battle*, 17.

80. Pyle, *Here Is Your War*, 151.

81. Standifer, *Not in Vain*, 84.

82. Walter Bernstein, *Keep Your Head Down* (New York: Viking, 1945), 149.

83. Burgett, *Seven Roads*, 77.

84. Harold J. Gordon, Jr., *One Man's War: A Memoir of World War II*, ed. Nancy M. Gordon (New York: Apex Press, 1999), 9.

85. Burgett, *Seven Roads*, 205.

86. Arnold, "'Easy' Memories," 29.

87. Paul Fussell, *Wartime: Understanding and Behavior in the Second World War* (New York: Oxford University Press, 1989), 282.

88. Phil Hannon in Astor, *Blood-Dimmed Tide*, 3.

89. Babcock, *Taught to Kill*, 60.

90. Harry Martin in Astor, *Blood-Dimmed Tide*, 90.

91. Babcock, *Taught to Kill*, 35.

92. Thomas Evans, "There Is No Use Trying to Dodge Shot," *Civil War Times Illustrated* 6 (January 1968): 45.

93. Frederick Heuring diary entry, 7 April 1862, Heuring Journal, IHS.

94. David Beem to fiancée, 14 July 1861, Beem Papers, IHS.

95. Josiah Atkins diary entry, 8 July 1781, in *The Diary of Josiah Atkins*, ed. Steven E. Kagle (New York: New York Times, 1975), 39–40.

96. Henry Welch to Polly and Franklin Tanner, 11 October 1862, Welch Papers, MHI.

97. Gordon, *One Man's War*, 17.

98. Eugene Schroeder in Drez, *Voices*, 187. The extent of this fatalism, and its utility in counteracting combat anxiety, is particularly difficult to gauge in twentieth-century soldiers, however. While such remarks appeared in numerous soldiers' remembrances, Dollard argued against the broad applicability of such beliefs in fatalism and luck, judging that they affected only about one quarter of the veterans in his sample. Dollard, *Fear in Battle*, 15.

99. David Beem to fiancée, July 1861, Beem Papers, IHS.

100. Arnold, "'Easy' Memories," 81.

101. Babcock, *Taught to Kill*, 62.

102. Arnold, "'Easy' Memories," 31.

103. Jack Keating in Drez, *Voices*, 278.

104. Even a movie as suffused with the "band of brothers" explanation for soldiers' motivation as *Saving Private Ryan* contains tantalizing glimpses of this mechanism. Tellingly, those glances occur in the heat of battle, not in the extended sequences between fights in which the soldiers muse on the justification for the war and for their unusual mission. One such scene occurs at the beginning of the film during the famous Omaha Beach sequence: struggling to cross the fire-raked beach, the film's Ranger captain protagonist pauses briefly and attempts to rally a crowd of terrified GIs who have taken refuge behind a bundle of girders employed as a tank obstacle. One panicked soldier justifies his refusal to go any further forward by noting that the meager obstacle is "the only thing between us and the Almighty." The captain urges the cowering soldiers to leave the safety of their temporary shelter not by explaining that the unit's mission is to clear space for successive waves of the assault or by urging them to support their comrades-in-arms, or even by issuing a direct order to advance. Rather, he makes a straightforward appeal to the soldiers' own sense of self-preservation: "Every inch of this beach," he shouts over the cacophony of battle, "has been pre-sighted by enemy fire." The obstacle, in other words, offers only the illusion of safety, and the logical conclusion is stark: "You stay here, you're a dead man."

105. S. L. A. Marshall, *Men against Fire: The Problem of Battle Command in Future War* (New York: William Morrow, 1947), 22.

Chapter 3. Training

1. Thomas Jacobs letter, probably to mother, undated, Thomas Jacobs Papers, U.S. Army Military History Institute, Carlisle, Pennsylvania (hereafter cited as MHI).

2. Paddy Griffith, *Rally Once Again: Battle Tactics of the American Civil War* (Marlborough, UK: Crowood Press, 1987), 105.

3. Bell Wiley, *The Life of Billy Yank, the Common Soldier of the Union* (Baton Rouge: Louisiana State University Press, 1952), 53; Griffith, *Rally Once Again*, 105–107.

4. Leon C. Standifer, *Not in Vain: A Rifleman Remembers World War II* (Baton Rouge: Louisiana State University Press, 1992), 46–90.

5. In "Training, Morale and Modern War," historian Hew Strachan argues that modern, twentieth-century training served an additional function: overcoming one of the most deeply rooted societal taboos by preparing soldiers to kill other humans. That discussion has given rise to an active debate among scholars. David Grossman typifies researchers on one side, arguing that the deeply ingrained human resistance to killing members of the species demanded elaborate

training rituals designed to circumvent a biologically hard-wired reluctance, while Joanna Bourke and others argue precisely the opposite, holding that many soldiers in fact reveled in the act of killing during wartime. David Grossman, *On Killing: The Psychological Cost of Learning to Kill in War and Society* (New York: Little, Brown, 1995); Joanna Bourke, *An Intimate History of Killing: Face-to-Face Killing in Twentieth-Century Warfare* (London: Granta Books, 1999). The importance of specific exercises in that dimension of warfare appeared particularly critical for soldiers of the twentieth century. Interestingly, explicit preparation to kill formed a relatively less important part of eighteenth- and nineteenth-century training—in part, no doubt, because of the difficulties of aimed fire and specific targeting in the battles of those eras. Just as soldiers were less likely to feel singled out while packed into the ranks, volley fire resembled a firing squad more than a sniper: judging the effects of any individual bullet proved far more difficult. Hew Strachan, "Training, Morale and Modern War," *Journal of Contemporary History* 42, no. 2 (2006): 211–227.

6. Thomas E. Rodgers's "Billy Yank and G.I. Joe: An Exploratory Essay on the Sociopolitical Dimensions of Soldier Motivation" argues that the differences between the societies from which Civil War and World War II soldiers sprang explain many of the differences between training in the two eras: nineteenth-century volunteers came to the army with an internalized "republican spirit" and conceptions of a public kind of masculine patriotism that enabled them to perform under fire. The more independent-minded soldiers of the mid-twentieth century, Rodgers argues, lacked the same degree of identification with the cause and required more elaborate and more formal training to prepare them for combat. Rodgers's analysis is compelling but discounts one critical and reasonable explanation for many of the differences in the nature of their training: the specific demands of the battlefields for which the two groups were being prepared differed profoundly. Thomas E. Rodgers, "Billy Yank and G.I. Joe: An Exploratory Essay on the Sociopolitical Dimensions of Soldier Motivation," *Journal of Military History* 69 (January 2005): 93–121.

7. James McPherson, *For Cause and Comrades: Why Men Fought in the Civil War* (New York: Oxford University Press, 1997), 46.

8. Reid Mitchell discusses the psychological transformation that accompanied the transition from civilians to soldiers in *Civil War Soldiers: Their Expectations and Their Experiences* (New York: Viking, 1988), 56–89; Bell Wiley covers some of the practical realities of Civil War training camps in *Billy Yank*, 49–54.

9. Grady McWhiney and Perry Jamieson, *Attack and Die: Civil War Military Tactics and the Southern Heritage* (Tuscaloosa: University of Alabama Press, 1982), 31.

10. Nelson Chapin to wife, 23 April 1862, Chapin Papers, MHI.

11. James Adams to sister, 8 November 1862, Adams Papers, Indiana Historical Society, Indianapolis, Indiana (hereafter cited as IHS).

12. George Rowland to brother, 26 August 1863, Rowland Papers, MHI.

13. Richard F. Ebbins to sister, 25 December 1862, Ebbins Papers, MHI.

14. Nelson Chapin to wife, 23 April 1862, Chapin Papers, MHI.

15. Sylvia R. Frey, *The British Soldier in North America: A Social History of Military Life in the Revolutionary Period* (Austin: University of Texas Press, 1981), 100.

16. In the early twentieth century, the phenomenon of increased performance of rigorously practiced tasks while under stress became known as the Yerkes-Dodson law, after the psychologists who studied and charted its influence. They described the correlation between performance of those overlearned tasks and stress with an inverted U-shaped curve. To a point, the addition of stresses actually improved performance; after a point, the presence of stress first eroded and then destroyed the ability to perform even overlearned tasks — an observation generations of soldiers in battle could affirm.

17. George Rowland to father, 6 August 1863, Rowland Papers, MHI.

18. Historian William McNeill refers to the phenomenon of increased group unity through constant communal repetition as "muscular bonding," and emphasizes its unique ability to knit the members of a group together. William H. McNeill, *Keeping Together in Time: Dance and Drill in Human History* (Cambridge, MA: Harvard University Press, 1995), 1–12.

19. John J. Sherman, undated letter, in McPherson, *Cause and Comrades*, 48.

20. Michael D. Doubler, *Closing with the Enemy: How GIs Fought the War in Europe, 1944–1945* (Lawrence: University Press of Kansas, 1994), 251–252; Peter R. Mansoor, *The GI Offensive in Europe: The Triumph of American Infantry Divisions, 1941–1945* (Lawrence: University Press of Kansas, 1999), 29–31.

21. John W. Baxter, *World War II Experiences of John W. Baxter*, unpublished manuscript, 99th Infantry Division Collection, MHI, 6.

22. Roscoe C. Blunt, *Foot Soldier: A Combat Infantryman's War in Europe* (Cambridge, MA: Da Capo Press, 2001), 15.

23. Baxter, *Experiences*, 7.

24. *How to Get Along in the Army, by "Old Sarge"* (New York: D. Appleton-Century, 1942), 54.

25. *Psychology for the Fighting Man, Prepared for the Fighting Man Himself by a Committee of the National Research Council with the Collaboration of Science Service as a Contribution to the War Effort* (Washington, DC: Infantry Journal, 1943), 200–201.

26. Standifer, *Not in Vain*, 7.

27. John Zmudzinski in Ronald J. Drez, ed., *Voices of D-Day: The Story of the Allied Invasion Told by Those Who Were There* (Baton Rouge: Louisiana State University Press, 1994), 239.

28. John B. Babcock, *Taught to Kill: An American Boy's War from the Ardennes to Berlin* (Washington, DC: Potomac Books, 2005), 48–49.

29. Dare Kibble in Chet Cunningham, *Hell Wouldn't Stop: An Oral History of the Battle of Wake Island* (New York: Carroll & Graf, 2002), 106–107.

30. Samuel Stouffer et al., *The American Soldier*, vol. 2, *Combat and Its Aftermath* (Princeton, NJ: Princeton University Press, 1949), 97–98.

31. Based on their interviews with veteran soldiers, Stouffer's team concluded that effective training could help minimize the disruptive effects of fear on the battlefield by teaching recruits "specific, appropriate reactions to be made to

various combat conditions." Exposure to realistic battle stimuli during their training could introduce and condition those reactions. Stouffer, *American Soldier*, 193.

32. Oliver Willcox Norton to sister, 19 December 1861, in *Army Letters, 1861–1865* (Chicago: O. L. Deming, 1903), 37–38.

33. John Dollard, *Fear in Battle* (New Haven, CT: Institute of Human Relations, 1943), 14.

34. Dollard's interviewees, who believed they had suffered more casualties in early actions as a result of ineffective training, were "all the more keenly aware of the importance of careful preparation." Dollard, *Fear in Battle*, 14.

35. Baxter, *Experiences*, 7.

36. Dollard's Spanish Civil War veterans expressed near unanimity on this point: fully 99% thought that having a veteran explain to the men on the basis of personal experience the life-saving importance of the things they were learning made them better soldiers. Dollard, *Fear in Battle*, 15.

37. Ralph Lewis, "Officer–Enlisted Men's Relationships," *American Journal of Sociology* 52, no. 5 (March 1947): 413.

38. Standifer, *Not in Vain*, 51.

39. Baxter, *Experiences*, 7.

40. Gail Beccue in Drez, *Voices*, 275.

41. Baxter, *Experiences*, 7. Historian and veteran NCO Robert Rush suggests another benefit to using live rounds during training in one of the notes to *Hell in Hürtgen Forest*, arguing that "there is nothing like live ammunition during a training exercise to get the heart pumping. Because of the high stress involved, soldiers learn quickly." Robert S. Rush, *Hell in Hürtgen Forest: The Ordeal and Triumph of an American Infantry Regiment* (Lawrence: University Press of Kansas, 2001), 79.

42. *Old Sarge*, 71.

43. Marion Hargrove, *See Here, Private Hargrove* (New York: Henry Holt, 1942), 4–5.

44. *Old Sarge*, 94.

45. Ibid., v.

46. Lee Kennett, *G.I.: The American Soldier in World War II* (Norman: University of Oklahoma Press, 1987), 48.

47. Gerald Linderman discusses the assumption among many recruits that "the consequences of combat would be determined by the effectiveness of the soldier's own reasoning and of his consequent actions," in *The World within War: America's Combat Experience in World War II* (New York: Free Press, 1997), 8–9.

48. Edward C. Arn, *Arn's War: Memoirs of a World War II Infantryman, 1940–1946*, ed. Jerome Mushkat (Akron: University of Akron Press, 2006), 43.

49. Baxter, *Experiences*, 7.

50. Ibid., 7–8.

51. Standifer, *Not in Vain*, 52–53.

52. Jim Bowers Questionnaire, 99th Infantry Division Collection, MHI.

53. Standifer, *Not in Vain*, 53.

54. Ibid., 73. To emphasize that very point, another mock tombstone at Fort

Benning carried the epitaph "Here lie the bones of Lieutenant Jones / A graduate of this institution / He died on the night of his very first fight / While using the school solution." Doubler, *Closing with the Enemy*, 265.
55. Standifer, *Not in Vain*, 52.
56. Ibid., 72.
57. Fred Gole Questionnaire, 99th Division Collection, MHI.
58. Blunt, *Foot Soldier*, 31.
59. Babcock, *Taught to Kill*, 58.
60. Harry S. Arnold, "'Easy' Memories: The Way It Was," unpublished manuscript, 99th Division Collection, MHI, 34.
61. Blunt, *Foot Soldier*, 2.

Chapter 4. Leadership

1. Aldace Freeman Walker to father, 18 September 1864, in *Quite Ready to Be Sent Somewhere: The Civil War Letters of Aldace Freeman Walker*, ed. Tom Ledoux (Victoria, BC: Trafford Publishing, 2002), 300.
2. William Manchester, *Goodbye, Darkness: A Memoir of the Pacific War* (Boston: Little, Brown, 1979), 3–7.
3. As with the changing nature of courage in battle, the requirements of leading troops in combat have given rise to an impressive literature of their own. Besides the scores of pamphlets and studies produced by the army itself over the past century and a half, a number of scholars have described and analyzed the way leaders prepared for battle and the tools they employed to encourage their charges once combat began. Robert Middlekauff's article "Why Men Fought in the American Revolution" reflects in passing on the influence provided by officers in both the British and American armies during that conflict. Don Higginbotham speaks more directly to the direction provided by officers in the Continental Army (both in and out of combat) in *The War of American Independence*, and he discusses the composition, attitudes, traditions, and influences of military leadership during that conflict in *War and Society in Revolutionary America*. Charles Royster examines the influences of leadership, especially as a motivator that sustained troops during the hardships of campaign, in *A Revolutionary People at War*. Civil War scholars have reflected on the nature of combat leadership, from Gerald Linderman in *Embattled Courage* (who distinguished between "conservator" commanders and "destroyer" commanders, according to their willingness to risk the lives of those in their commands); James I. Robertson, Jr., describes the influence officers exerted in combat (and before combat, in their addresses to men preparing to face fire) in *Soldiers Blue and Gray*; James McPherson reflects on the connection between combat leadership and combat motivation in the armies of the American Civil War in *For Cause and Comrades*; Peter Kindsvatter reflects on the influence of combat leadership on the behavior of American infantry soldiers in the conflicts of the twentieth century in *American Soldiers*. Robert Middlekauff, "Why Men Fought in the American Revolution," *Huntington Library Quarterly* 43 (Spring 1980): 135–148;

Don Higginbotham, *The War of American Independence: Military Attitudes, Policies, and Practice, 1763–1789* (New York: Macmillan, 1971), 398–413; Higginbotham, *War and Society in Revolutionary America: The Wider Dimensions of Conflict* (Columbia: University of South Carolina Press, 1988), 84–105; Charles Royster, *A Revolutionary People at War: The Continental Army and American Character, 1775–1783* (Chapel Hill: University of North Carolina Press, 1979); Gerald Linderman, *Embattled Courage: The Experience of Combat in the American Civil War* (New York: Free Press, 1987), 201–210; James I. Robertson, Jr., *Soldiers Blue and Gray* (Columbia: University of South Carolina Press, 1988), 215–216; James McPherson, *For Cause and Comrades: Why Men Fought in the Civil War* (New York: Oxford University Press, 1997), 58–61; Peter Kindsvatter, *American Soldiers: Ground Combat in the World Wars, Korea, and Vietnam* (Lawrence: University Press of Kansas, 2003), 229–245.

4. *Psychology for the Fighting Man, Prepared for the Fighting Man Himself by a Committee of the National Research Council with the Collaboration of Science Service as a Contribution to the War Effort* (Washington, DC: Infantry Journal, 1943), 200.

5. Lester Atwell, *Private* (New York: Simon & Schuster, 1958), 31. The film *Saving Private Ryan* depicts a similar lesson during an exchange early in the film. Moments before the ramp on their landing craft drops on Omaha Beach, a sergeant (presented as a hardened veteran of the battle at Kasserine Pass and the landings at Anzio) attempts to steady nervous troops with an analogous message. "I want to see plenty of space between men," he shouts over the noise of the boat's engines. "Three men is a juicy opportunity" for enemy gunners, he reminds them, before assuring them that "one man is a waste of ammo." The lesson, with its greatly oversimplified reassurances, offers further reinforcement for the soldiers' sense that the chaos of battle was nevertheless in some ways under their control: *do this and you'll be okay.*

6. Frank Raila in Gerald Astor, ed., *A Blood-Dimmed Tide: The Battle of the Bulge by the Men Who Fought It* (New York: Donald I. Fine, 1992), 194.

7. William Monsanto Questionnaire, 99th Division Collection, U.S. Army Military History Institute, Carlisle, Pennsylvania (hereafter cited as MHI).

8. Ernie Pyle, *Here Is Your War* (New York: Henry Holt, 1943), 160.

9. Robert Adams in Ronald J. Drez, ed., *Voices of D-Day: The Story of the Allied Invasion Told by Those Who Were There* (Baton Rouge: Louisiana State University Press, 1994), 230–231.

10. Lt. Col. Edward Lyman Munson, Jr., *Leadership for American Army Leaders* (Washington, DC: Infantry Journal, 1943), 1.

11. Jeremiah Preston pension application, 20 December 1843, in Lawrence E. Babits, *Devil of a Whipping: The Battle of Cowpens* (Chapel Hill: University of North Carolina Press, 1998), 103.

12. Henry Wells pension application, 29 January 1834, in Babits, *Devil of a Whipping*, 100.

13. In Lyman Copeland Draper et al., *King's Mountain and Its Heroes: A History of the Battle of King's Mountain, October 7th, 1780, and the Events Which Led to It* (Cincinnati: P.G. Thomson, 1881), 285–286.

14. Royster, *Revolutionary People at War*, 220.

15. Thomas Young, "Memoir of Major Thomas Young," *Orion Magazine* (October/November 1843): 88.

16. Ibid., 100.

17. In Draper et al., *King's Mountain*, 285–286.

18. Robert Carter to father, July 1861, *Four Brothers in Blue: Or, Sunshine and Shadows in the War of the Rebellion* (Austin: University of Texas Press, 1978), 9.

19. Rice C. Bull, *Soldiering: The Civil War Diary of Rice C. Bull, 123rd New York Volunteer Infantry*, ed. Jack K. Bauer (San Rafael, CA: Presidio Press, 1977), 118.

20. George Rowland to father, 19 November 1863, Rowland Papers, MHI.

21. Victor E. Comte in Earl J. Hess, *The Union Soldier in Battle: Enduring the Ordeal of Combat* (Lawrence: University Press of Kansas, 1997), 121.

22. Aldace Walker to father, 18 September 1864, in Walker, *Quite Ready*, 300.

23. Aldace Walker to father, 21 September 1864, in ibid., 304.

24. Emory Upton, *A New System of Infantry Tactics, Double and Single Rank, Adapted to American Topography and Improved Fire-Arms* (New York: Appleton, 1867), 111.

25. George Rowland to brother, 12 August 1863, George Rowland Papers, MHI.

26. Joshua K. Callaway to wife, 17 September 1863, in Callaway, *The Civil War Letters of Joshua K. Callaway*, ed. Judith Lee Hallock (Athens: University of Georgia Press, 1997), 133.

27. Lorenzo Vanderhoef diary entry, 21 November 1861, in Kenneth R. Martin and Ralph Linwood Snow, eds., *"I Am Now a Soldier!": The Civil War Diaries of Lorenzo Vanderhoef* (Bath, ME: Patten Free Library, 1990), 66.

28. Young, "Memoir of Major Thomas Young."

29. Henry Welch to Polly and Franklin Tanner, 24 May 1863, Welch Papers, MHI.

30. *Harper's Weekly* 4, 17 January 1863, 75–78.

31. In Samuel Stouffer et al., *The American Soldier*, vol. 2, *Combat and Its Aftermath* (Princeton, NJ: Princeton University Press, 1949), 117.

32. Major James A. Moss, *Officers' Manual* (Menasha, WI: George Banta, 1917), 422–423.

33. Stouffer, *American Soldier*, 2:102.

34. Munson, *Leadership for American Army Leaders*, 1. The *Psychology for the Fighting Man* booklet included an identically titled chapter, whose opening paragraph insisted, "There are no born leaders. All leadership is based on learning how to deal with men." *Psychology for the Fighting Man*, 382.

35. Stouffer, *American Soldier*, 2:102.

36. *Psychology for the Fighting Man*, 367.

37. Alvin Boeger Questionnaire, MHI.

38. Roscoe C. Blunt, Jr., *Foot Soldier: A Combat Infantryman's War in Europe* (Cambridge, MA: Da Capo Press, 2001), 148–149.

39. Thomas Sams Bishop diary entry, 18 November 1944, 99th Infantry Division Collection, MHI.

40. *Psychology for the Fighting Man*, 373–374.

41. *How to Get Along in the Army, by "Old Sarge"* (New York: D. Appleton-Century, 1942), 32.

42. Boeger Questionnaire, MHI.

43. Robert R. Palmer, Bell I. Wiley, and William R. Keast, *The United States Army in World War II: The Procurement and Training of Ground Troops* (Washington, DC: Historical Division, Department of the Army, 1948), 341–342.

44. William Thompson Lusk, *War Letters of William Thompson Lusk, Captain, Assistant Adjutant-General, United States Volunteers 1861–1863* (New York: Private printing, 1911), 247–248.

45. Stouffer, *American Soldier*, 2:117.

46. *Psychology for the Fighting Man*, 290.

47. Gerald Linderman, *The World within War: America's Combat Experience in World War II* (New York: Free Press, 1997), 17.

48. Field Manual 100-5, *Field Service Regulations* (Washington, DC: U.S. Government Printing Office, 1941), 19.

49. Jack Hartzog Questionnaire, 78th Infantry Division Collection, MHI.

50. John F. Campbell Questionnaire, 99th Infantry Division Collection, MHI.

51. Frank Raila in Astor, *Blood-Dimmed Tide*, 16.

52. Stouffer, *American Soldier*, 2:90.

53. John Dollard, *Fear in Battle* (New Haven, CT: Institute of Human Relations, 1943), 58.

54. *Psychology for the Fighting Man*, 383.

55. Boeger Questionnaire, MHI.

56. Joe Dine, *Fighting 36th Historical Quarterly* (Winter 1984); quoted in McManus, *Deadly Brotherhood*, 216.

57. Bill Mauldin, *Bill Mauldin's Army* (Novato, CA: Presidio Press, 1979), 98.

58. Pyle, *Here Is Your War*, 160.

59. Jason Byrd interview in McManus, *Deadly Brotherhood*, 207.

60. Pyle, *Here Is Your War*, 136.

61. Linderman, *Embattled Courage*, 272–273.

62. William T. Sherman, *Memoirs of General William T. Sherman* (New York: D. Appleton, 1875), 2:405.

63. Bowers Questionnaire, MHI.

64. Frank Raila in Astor, *Blood-Dimmed Tide*, 16.

65. Harold Sletten, World War II questionnaire #2111, MHI, in McManus, *Deadly Brotherhood*, 205.

66. Munson, *Leadership*, 6.

67. Ibid., 7.

68. Ibid., 9.

69. Campbell Questionnaire, 99th Infantry Division Collection, MHI.

70. Young, "Memoir of Major Thomas Young," 88.

71. Thomas Livermore, *Days and Events, 1860–1866* (Boston: Houghton Mifflin, 1920), 133.

72. Harry S. Arnold, "'Easy' Memories: The Way It Was," unpublished manuscript, 99th Division Collection, MHI, 71.

246 Notes to Pages 148–154

73. Carl Carthedge in Drez, *Voices*, 73.
74. John D. Boone in Drez, *Voices*, 127.
75. Leon C. Standifer, *Not in Vain: A Rifleman Remembers World War II* (Baton Rouge: Louisiana State University Press, 1992), 101.
76. Mauldin, *Bill Mauldin's Army*, 303.
77. David Hunter Strother diary entry, 25 March 1862, in David Hunter Strother, *A Virginia Yankee in the Civil War: The Diaries of David Hunter Strother*, ed. Cecil D. Edby, Jr. (Chapel Hill: University of North Carolina Press, 1961), 21.

Chapter 5. Weaponry

1. Robert Walker and Tommy Horne in Ronald J. Drez, ed., *Voices of D-Day: The Story of the Allied Invasion Told by Those Who Were There* (Baton Rouge: Louisiana State University Press, 1994), 223, 128.
2. Donald Burgett, *Seven Roads to Hell: A Screaming Eagle at Bastogne* (Novato, CA: Presidio Press, 1999), 22. Modern psychological studies of the elements affecting the morale of the individual soldier suggest that only a very small group of factors affect troops' willingness to enter and remain in combat. Human qualities constitute most of the list: confidence in one's own skills as a soldier, the nature of relationships with commanders, and trust in comrades. The other critical factor that emerged was the soldier's trust in his weapons. Gregory Belenky, Shabtai Noy, and Zahava Solomon, "Battle Stress, Morale, Cohesion, Combat Effectiveness, Heroism, and Psychiatric Casualties: The Israeli Experience," in Gregory Belenky, ed., *Contemporary Studies in Combat Psychiatry* (New York: Greenwood Press, 1987), 15.
3. Robert Knauss Questionnaire, 37th Infantry Division Collection, U.S. Army Military History Institute, Carlisle, Pennsylvania (hereafter cited as MHI).
4. S. H. M. Byers, "How Men Feel in Battle: Recollections of a Private at Champion Hills," *Annals of Iowa* 2 (July 1896): 443.
5. Belenky, Noy, and Solomon, "Battle Stress," 15.
6. A. Z. Adkins, Jr., and Andrew Z. Adkins, III, *You Can't Get Much Closer Than This: Combat with Company H, 317th Infantry Regiment, 80th Division* (Havertown, PA: Casemate Publishing, 2005), 18–19.
7. Jim Foley in Gerald Astor, *A Blood-Dimmed Tide: The Battle of the Bulge by the Men Who Fought It* (New York: Donald I. Fine, 1992), 119–120.
8. Thomas Valence in Drez, *Voices*, 202.
9. Lorenzo Vanderhoef diary entry, 5 July 1861, in Vanderhoef, *"I Am Now a Soldier!": The Civil War Diaries of Lorenzo Vanderhoef*, ed. Kenneth R. Martin and Ralph Linwood Snow (Bath, ME: Patten Free Library, 1990), 34.
10. Bell I. Wiley, *The Life of Johnny Reb, the Common Soldier of the Confederacy* (New York: Bobbs Merrill, 1943), 290–291.
11. Jonas Denton Elliot to wife, 15 September 1863, Elliot Papers, MHI.
12. Joshua K. Callaway to wife, 2 June 1862, in Callaway, *The Civil War Letters of Joshua K. Callaway*, ed. Judith Lee Hallock (Athens: University of Georgia Press, 1997), 25.

13. Ross S. Carter, *Those Devils in Baggy Pants* (New York: Appleton-Century-Crofts, 1951), 14.

14. Leon C. Standifer, *Not in Vain: A Rifleman Remembers World War II* (Baton Rouge: Louisiana State University Press, 1992), 119.

15. John B. Babcock, *Taught to Kill: An American Boy's War from the Ardennes to Berlin* (Washington, DC: Potomac Books, 2005), 14–15.

16. Bell I. Wiley, *The Life of Billy Yank, the Common Soldier of the Union* (New York: Bobbs Merrill, 1951), 66–67.

17. Some crude repeating weapons were available during the Civil War, but both Union and Confederate leaders rejected the idea of issuing them to the infantry in the ranks, since ammunition was in such short supply. Commanders feared that soldiers would simply shoot through their ammunition in a few moments and then, with nothing left to fire at the enemy and no means to respond to attacks, flee the battlefield. Given the limited amount of ammunition the individual soldier could carry into combat, the lengthier reloading times of those Springfield rifles appeared to help, in some way, to keep the unit together.

18. Charles Royster, *A Revolutionary People at War: The Continental Army and American Character, 1775–1783* (Chapel Hill: University of North Carolina Press, 1979), 35.

19. Wiley, *Johnny Reb*, 293.

20. Harry S. Arnold, "'Easy' Memories: The Way It Was," unpublished manuscript, 99th Division Collection, MHI, 72.

21. Babcock, *Taught to Kill*, 47–48.

22. Sylvia R. Frey, *The British Soldier in North America: A Social History of Military Life in the Revolutionary Period* (Austin: University of Texas Press, 1981), 28.

23. Lester Atwell, *Private* (New York: Simon & Schuster, 1958), 7–8.

24. Richard Lovett in John C. McManus, *The Deadly Brotherhood: The American Combat Soldier in World War II* (Novato, CA: Presidio Press, 1998), 44.

25. Gil Murdoch, in Drez, *Voices*, 204.

26. George Hanger, *Colonel George Hanger's Advice to All Sportsmen* (London: J. J. Stockdale, 1814), 144.

27. Grady McWhiney and Perry Jamieson, *Attack and Die: Civil War Military Tactics and the Southern Heritage* (Tuscaloosa: University of Alabama Press, 1982), 49.

28. Philip M. Cole, *Civil War Artillery at Gettysburg: Organization, Equipment, Ammunition and Tactics* (Cambridge, MA: Da Capo Press, 2002), 227–231.

29. George H. Elsbury to Colt & Co., 3 September 1863, Elsbury Letter, MHI.

30. Donald Greener in McManus, *Deadly Brotherhood*, 39–40.

31. Roscoe C. Blunt, Jr., *Foot Soldier: A Combat Infantryman's War in Europe* (Cambridge, MA: Da Capo Press, 2001), 179.

32. Robert Walker in Drez, *Voices*, 222.

33. Phil Hannon in Astor, *Blood-Dimmed Tide*, 165.

34. Gordon C. Rhea, *The Battle of the Wilderness, May 5–6, 1864* (Baton Rouge: Louisiana State University Press, 1994), 171.

35. Atwell, *Private*, 44–45.

36. Harry Parley in Drez, *Voices*, 210.

37. Demetrius Paris in Astor, *Blood-Dimmed Tide*, 157.

38. Gene Curry in McManus, *Deadly Brotherhood*, 39.

39. The modern soldier's admonition, *Tracers point both ways*, originated with the fact that those streaks were said to track back to their source.

40. Ernie Pyle, *Here Is Your War* (New York: Henry Holt, 1943), 17 .

41. Babcock, *Taught to Kill*, 45.

42. Walter Powell in McManus, *Deadly Brotherhood*, 46.

43. *Fighting 36th Historical Quarterly* (Spring 1992): 30, in McManus, *Deadly Brotherhood*, 46.

44. Arnold, "'Easy' Memories," 34.

45. Ebeneezer Hannaford, "In the Ranks at Stones River," in *Battles a nd Leaders of the Civil War*, ed. Peter Cozzens, vol. 6 (Champaign: University of Illinois Press, 2004), 182.

46. Such a symbiotic connection between soldiers and their weapons was common in other armies fighting in the dispersed system. During the battle of Stalingrad, for example, the slogan of the Soviet 62nd Army instructed soldiers of the Red Army to "look after your weapon as carefully as your eyes." Anthony Beevor, *Stalingrad: The Fateful Siege, 1942–1943* (New York: Viking, 1998), 154.

47. Bill Mauldin, *Bill Mauldin's Army* (Novato, CA: Presidio Press, 1979), 11 1.

48. The new twentieth-century attitude toward the pairing of soldier and weapon appears in a memorable scene in Stanley Kubrick's 1987 Vietnam War film *Full Metal Jacket*, in which a group of Marine recruits parade through their barracks, rifles on shoulders, while intoning in unison: "Without me, my rifle is useless; without my rifle, I am useless."

49. Burgett, *Seven Roads to Hell*, 21.

50. Blunt, *Foot Soldier*, 131.

51. Colin McLaurin, *The Twenty-Niner Newsletter* (July 1995); Thomas Rosell interview with Dr. Charles W. Johnson, 17 August 1993, World War II Veterans Project, Special Collections, University of Tennessee, Knoxville, in McManus, *Deadly Brotherhood*, 49.

52. Harold J. Gordon, Jr., *One Man's War: A Memoir of World War II*, ed. Nancy M. Gordon (New York: Apex Press, 1999), 11.

53. Adkins and Adkins, *Can't Get Much Closer*, 60.

54. Bill Davidson, *Cut Off: Behind Enemy Lines in the Battle of the Bulge with Two Small Children, Ernest Hemingway, and Other Assorted Misanthropes* (New York: Stein and Day, 1972), 12.

55. Walter Bernstein, *Keep Your Head Down* (New York: Viking, 1945), 111.

56. Paul Fussell in Richard M. Stannard, *Infantry: An Oral History of a World War II American Infantry Battalion* (New York: Twayne, 1993), 101.

57. Henry Welch to father, 6 July 1864, Welch Papers, MHI.

58. Standifer, *Not in Vain*, 96.

59. Burgett, *Seven Roads*, 40–41.

60. *How to Get Along in the Army, by "Old Sarge"* (New York: D. Appleton-Century, 1942), 64.

61. S. L. A. Marshall, *Men against Fire: The Problem of Battle Command in Future War* (New York: William Morrow, 1947), 43.

Chapter 6. Comradeship

1. Joshua Lawrence Chamberlain, *Passing of the Armies: An Account of the Final Campaign of the Army of the Potomac Based upon Personal Reminiscences of the Fifth Army Corps* (New York: G. P. Putnam's Sons, 1915), 20. Usually interpreted as evidence of the power of bonds of affection to sustain troops in battle, the second half of Chamberlain's assertion can also be read as testimony to the effectiveness of modeled behavior in combat.

2. Joseph Dougherty Questionnaire, 99th Division Collection, U.S. Army Military History Institute, Carlisle, Pennsylvania (hereafter cited as MHI).

3. S. L. A. Marshall, *Men against Fire: The Problem of Command in Future War* (New York: William Morrow, 1947), 42. In the years following the publication of *Men against Fire*, many of Marshall's findings (particularly the remarkable assertion that in battle, fewer than 25 percent of a unit's riflemen actually fired their weapons) appeared in scores of important works of military history. In the last quarter-century, however, scholars have called Marshall's methods, statistics, and conclusions into question. Marshall claimed to have based many of his findings on more than 400 after-action interviews with rifle companies, but investigation by researchers in the mid-1980s suggested that Marshall had fabricated those interviews (and the statistics gleaned from them) partially if not wholly. Those realizations dramatically undermined faith in Marshall's more controversial assertions, particularly those dealing with fire ratios. Marshall's broader points about the importance of small-group cohesion, however, have not suffered the same degree of dispute. Peter Mansoor, *The GI Offensive in Europe: The Triumph of American Infantry Divisions, 1941–1945* (Lawrence: University Press of Kansas, 1999), 257–262; Roger J. Spiller, "S. L. A. Marshall and the Ratio of Fire," *RUSI Journal* 133, no. 4 (Winter 1988): 63–71.

4. Jonathan Shay, *Achilles in Vietnam: Combat Trauma and the Undoing of Character* (New York: Simon & Schuster, 1994), 40.

5. Marshall, *Men against Fire*; Edmund Shils and Morris Janowitz, "Cohesion and Disintegration in the Wehrmacht in World War II," *Public Opinion Quarterly* 12, no. 2 (Summer 1948): 280–315; Samuel Stouffer et al., *The American Soldier* (Princeton, NJ: Princeton University Press, 1949).

6. The same trope appeared in Ridley Scott's 2001 film about the American mission to retrieve survivors of a helicopter crash in Somalia, *Black Hawk Down*. Reflecting that he cannot satisfactorily convey to civilians at home why he chooses to go into combat, an elite U.S. soldier ascribes his willingness to go back into harm's way to the fact that there are still American soldiers at the crash site in

need of rescue: "When I go home, people say . . . 'Why do you do it, man? Why? Are you some kind of war junkie?' I won't say a goddamn word. Why? 'Cause they won't understand. They won't understand why we do it. They won't understand it's about the men next to you. And that's it. That's all it is."

7. Omer Bartov, *Hitler's Army: Soldiers, Nazis, and War in the Third Reich* (New York: Oxford University Press, 1991), 29–58.

8. Bartov devoted an entire chapter, "The Destruction of the Primary Group," to this finding and its implications, concluding that there was no clear connection between strong group cohesion and a unit's willingness to fight: though "the 'primary groups' did more or less disappear, the army fought with far greater determination and against far greater odds than at any other time in the past." Bartov, *Hitler's* Army, 33.

9. Robert S. Rush, *Hell in Hürtgen Forest: The Ordeal and Triumph of an American Infantry Regiment* (Lawrence: University Press of Kansas, 2001), 332–333.

10. Joseph T. Glatthaar, *General Lee's Army: From Victory to Collapse* (New York: Free Press, 2008), 315–317.

11. Ernie Pyle, *Here Is Your War* (New York: Henry Holt, 1943), 260. Robin Williams, Jr., one of the psychologists who worked on the original *American Soldier*, echoed this feeling of transience in American units of the Second World War: "One could not stay with a front-line combat organization over periods of months without receiving indelible impressions of the impermanence of specific persons"; Robin M. Williams, Jr., "Field Observations and Surveys in Combat Zones," *Social Psychology Quarterly* 47, no. 2 (1987): 186.

12. Philip Zimbardo, *The Lucifer Effect: Understanding How Good People Turn Evil* (New York: Random House, 2007).

13. Christian G. Appy, *Working-Class War: American Combat Soldiers and Vietnam* (Chapel Hill: University of North Carolina Press, 1993), 246–247. Lamont Steptoe, a Vietnam-era GI interviewed by historian Richard R. Moser, captured perfectly the way a unit's cohesive bonds could coalesce in this manner, even across previously divisive racial lines. Speaking of an unpopular leader, Steptoe remembered how "we banded together, blacks and white, we banded together and said, 'He's got to go.'" Richard R. Moser, *The New Winter Soldiers: GI and Veteran Dissent during the Vietnam Era* (New Brunswick, NJ: Rutgers University Press, 1996), 49.

14. Charles C. Moskos, Jr., "The American Combat Solider in Vietnam," *Social Issues* 31 (1975): 35.

15. Roger Little, "Buddy Relations and Combat Performance," in Morris Janowitz, ed., *The New Military: Changing Patterns of Organization* (New York: Russell Sage Foundation, 1964), 201.

16. While far less common than the "band of brothers" trope, this sentiment appears occasionally in Hollywood. The 1949 film *Twelve O'Clock High*, for example, features a scene in which the commander of an Eighth Air Force bomber group confronts a pilot who elected to break formation during a bombing run in order to go to the aid of a crippled plane. Upon learning that the pilot of the damaged bomber was the breakaway pilot's roommate, the commander admonishes the assembled crews that such friendships cannot interfere with their actions in

battle: "this group, this group, this group" must be their only obligation and their only loyalty, "their only reason for being," he reprimands them, moments before ordering the adjutant to shuffle room assignments to weaken the bonds of friendship.

17. Robert MacCoun, Elizabeth Kier, and Aaron Belkin, "Does Social Cohesion Determine Motivation in Combat? An Old Question with an Old Answer," *Armed Forces and Society* 32 (2006): 647. It is difficult to dismiss an intuitive, commonsense explanation for the willingness of soldiers to accept the risk of battle, particularly one endorsed by so many firsthand participants. Scholars who discount the ability of emotional bonds to motivate troops in combat note that people are notoriously unable to reliably perceive and report on the causes of their behavior. Often, these reasons occur below the level of cognition, where subjects themselves are only vaguely aware of them. Because these behaviors take place beneath conscious awareness, people become highly suggestible when it comes to explaining them: rather than reflecting on the causes of their behaviors, subjects often repeat what they understand to be the commonsense explanation for that behavior. In the case of primary group cohesion and combat motivation, what develops is a kind of feedback loop: because they are told so regularly by movies, documentaries, and books that soldiers generally fight for one another, combat veterans often repeat these well-known explanations when asked about their own reasons for fighting—and these statements in turn become part of the ever-growing body of anecdotal evidence supporting the relationship between cohesion and combat motivation. Robert J. MacCoun, "What Is Known about Unit Cohesion and Military Performance," in *Sexual Orientation and U.S. Military Personnel Policy: Options and Assessment* (Santa Monica, CA: RAND, 1993).

18. Marshall, *Men against Fire*, 22.

19. MacCoun, "What Is Known," 291.

20. Charles C. Moskos, Jr., *The American Enlisted Man: The Rank and File in Today's Military* (New York: Russell Sage Foundation, 1970), 145–147.

21. Shay, *Achilles in Vietnam*, 40.

22. In his 1981 article "The Social History of the American Soldier," historian Richard H. Kohn suggested that while the phenomenon of cohesion among primary groups has been more or less universal among American soldiers, "the literature on primary group cohesion has never clearly shown whether solidarity with the group acted as a prop to bolster men to endure the stress or as a motivation to carry out the mission and perform effectively in battle—or both." In his analysis of twentieth-century U.S. fighting men, *American Soldiers*, Peter Kindsvatter argued the latter case, holding that group loyalty was "both a significant prop and motivator" for troops in wartime. Because of its focus on individual behavior during actual battle, the analysis in this chapter takes a slightly different tack. Without question, the psychological support provided by friendly soldiers was critically important in encouraging soldiers to withstand the isolation, anxiety, and deprivation of military service—a valuable, if not indispensible, prop for soldiers. But that support functioned most noticeably between firefights: that is, as a sustaining motivator, in John Lynn's three-part conception of military

motivation. Within the maelstrom of combat, the support of comrades was at least as important in the way it appeared to heighten the survival of the individual soldier. Richard H. Kohn, "The Social History of the American Soldier: A Review and Prospectus for Research," *American Historical Review* 86, no. 3 (June 1981): 561; Peter S. Kindsvatter, *American Soldiers: Ground Combat in the World Wars, Korea, and Vietnam* (Lawrence: University Press of Kansas, 2003), 124.

23. Jonathan P. Stowe to friends, 11 July 1862, Stowe Papers, MHI.

24. Stephen Crane, *The Red Badge of Courage* (London: Folio Society, 1951), 92.

25. Henry Welch to Polly and Franklin Tanner, 10 September 1863, Welch Papers, MHI.

26. Crane, *Red Badge*, 47.

27. Garrett Watts pension application, in John C. Dann, *The Revolution Remembered: Eyewitness Accounts of the War for Independence* (Chicago: University of Chicago Press, 1980), 195.

28. Aldace Freeman Walker to father, 23 November 1862, in Walker, *Quite Ready to be Sent Somewhere: The Civil War Letters of Aldace Freeman Walker*, ed. Tom Ledoux (Victoria, BC: Trafford, 2002), 54.

29. Earl Hess presents a detailed treatment of this phenomenon in the Union Army in *The Union Soldier in Battle: Enduring the Ordeal of Combat* (Lawrence: University of Kansas Press, 1997), 88–89.

30. Henry Welch to Polly and Franklin Tanner, 15 July 1863, Welch Papers, MHI.

31. Charles T. Morey diary entry, 17 November 1863, Morey Papers, MHI.

32. Josiah Atkins diary entry, 6 July 1781, in Atkins, *The Diary of Josiah Atkins*, ed. Steven E. Kagle (New York: New York Times, 1975), 37–38.

33. Henry Welch to father, 23 September 1863, Welch Papers, MHI.

34. Rice C. Bull, *Soldiering: The Civil War Diary of Rice C. Bull, 123rd New York Volunteer Infantry*, ed. Jack K. Bauer (San Rafael, CA: Presidio Press, 1977), 164.

35. Henry Welch to Polly and Franklin Tanner, 9 June 1864, Welch Papers, MHI.

36. Abner Dunham to parents, 21 December 1863, in Abner Dunham, "Civil War Letters of Abner Dunham," ed. Mildred Thorne, *Iowa Journal of History* 53 (1955): 320–321.

37. Walter Bernstein, *Keep Your Head Down* (New York: Viking, 1945), 11.

38. David Hunter Strother diary entry, 19 March 1862, in David Hunter Strother, *A Virginia Yankee in the Civil War: The Diaries of David Hunter Strother*, ed. Cecil D. Eby, Jr. (Chapel Hill: University of North Carolina Press, 1961), 17.

39. In the confusion of battle, for example, Confederate general James Longstreet was shot by rebel troops at the Wilderness and did not return to combat for five months. James McPherson, *Battle Cry of Freedom: The Civil War Era* (New York: Oxford University Press, 1988), 725–726. Perhaps the best-known fratricidal incident of the Civil War, Thomas Jackson's mortal wound from friendly pickets while reconnoitering after the battle at Chancellorsville in 1863, is slightly different in that it occurred during nightfall, after the day's fighting had ended.

40. Harry S. Arnold, "'Easy' Memories: The Way It Was," unpublished manuscript, 99th Infantry Division Collection, MHI, 47.

41. Quoted in Stouffer, *American Soldier*, 2:99.

42. Henry Giles, *The GI Journal of Sergeant Giles* (Boston: Houghton Mifflin, 1965), 130–131.

43. Ross S. Carter, *Those Devils in Baggy Pants* (New York: Appleton-Century-Crofts, 1951), 134.

44. Quoted in Stouffer, *American Soldier*, 2:99. Roger Little, studying the infantry soldiers of the Korean conflict, built on these observations. As one of Little's subjects indicated, making and keeping friends was a matter of necessity for soldiers in the team-oriented system of linear tactics: "You've got to make every man in the squad your buddy to get things done." One's own safety depended in part upon the mutual faith comrades had in each other. "You've got to get down and work with them," the Korean-era veteran indicated, "and get them to feel that they can depend on you to stick by them." The same veteran cautioned against showing preferential treatment to one particular buddy, which might dissuade others in the unit from coming to the aid of a soldier in battle. Little, "Buddy Relations," 201.

45. Leon C. Standifer, *Not in Vain: A Rifleman Remembers World War II* (Baton Rouge: Louisiana State University Press, 1992), 52.

46. Lester Atwell, *Private* (New York: Simon & Schuster, 1958), 40.

47. Of course, it was not always impossible for armies in the dispersed system to use this kind of direct coercion. Numerous examples from the First and Second World Wars depict junior officers using immediate threats to prod their soldiers forward (one British officer from the First World War memorably threatened a private whose nerve failed by telling him, "You are a fucking coward and you will go to the trenches. I give fuck all for my life and I give fuck all for yours and I'll get you fucking well shot"). In Ben Shepherd, *A War of Nerves: Soldiers and Psychiatrists in the Twentieth Century* (Cambridge, MA: Harvard University Press, 2001), 68.

48. In his book on the American experience in World War II, Gerald Linderman suggested that comradeship "pushes as well as pulls" to get men into battle. Gerald Linderman, *The World within War: America's Combat Experience in World War II* (New York: Free Press, 1997), 266.

49. *How to Get Along in the Army, by "Old Sarge"* (New York: D. Appleton-Century, 1942), 10.

50. *Old Sarge*, 74, 87.

51. Roscoe C. Blunt, *Foot Soldier: A Combat Infantryman's War in Europe* (Cambridge, MA: Da Capo Press, 2001), 34.

52. Harry Martin, Jr., in Gerald Astor, *A Blood-Dimmed Tide: The Battle of the Bulge by the Men Who Fought It* (New York: Donald I. Fine, 1992), 8.

53. Donald Burgett, *Seven Roads to Hell: A Screaming Eagle at Bastogne* (Novato, CA: Presidio Press, 1999), 12.

54. Standifer, *Not in Vain*, 102–103.

55. Boeger Questionnaire, 99th Division Collection, MHI.

56. Burgett, *Seven Roads*, 49.

57. Ibid., 23–24.

58. Standifer, *Not in Vain*, 100.

59. Peter Richmond, *My Father's War: A Son's Journey* (New York: Simon & Schuster, 1996), 252–253.

60. Arnold, "'Easy' Memories," 46.

61. Standifer, *Not in Vain*, 108.

62. Burgett, *Seven Roads*, 3.

63. Bill Mauldin, *Bill Mauldin's Army* (Novato, CA: Presidio Press, 1979), 158.

64. Arnold, "'Easy' Memories," 106.

65. Standifer, *Not in Vain*, 100, 117.

66. Burgett, *Seven Roads*, 15.

67. Carter, *Devils*, 163–164.

68. Only months later did the soldier acquire a heightened appreciation for the interconnectedness of the efforts of *all* the branches: the infantrymen's derisive chant "was going to cost Patton the fuel he needed for the drive across France. It was going to give me frozen feet. A lot of kids were going to lose their feet because we were saying that the truck driver wasn't important." The "Blue Star Commandos," the soldier surmised, had found people in France who would respect him, "if he sold them some fuel or warm shoes." Standifer, *Not in Vain*, 106–107.

69. Carter, *Devils*, 193.

70. Ibid., 159.

71. Standifer, *Not in Vain*, 71–72.

72. Arnold, "'Easy' Memories," 39.

73. Harold J. Gordon, Jr., *One Man's War: A Memoir of World War II*, ed. Nancy M. Gordon (New York: Apex Press, 1999), 12.

74. Richard Cassiday in Ronald J. Drez, ed., *Voices of D-Day: The Story of the Allied Invasion Told by Those Who Were There* (Baton Rouge: Louisiana State University Press, 1994), 178.

75. Charles Ardant du Picq, *Battle Studies: Ancient and Modern War* (Harrisburg, PA: Military Service Publishing, 1947), 22. In his study of the British army's struggle to adjust to the new tactical realities of industrial warfare, Tim Travers argued that prewar British officers divided generally into two camps: those who wanted more discipline, and those who wanted to "liberate the enthusiasm and willingness to fight of the individual." Tim Travers, *The Killing Ground: The British Army, the Western Front, and the Emergence of Modern Warfare 1900–1918* (London: Allen & Unwin, 1987), 37–55.

Conclusion

1. S. H. M. Byers, *With Fire and Sword* (New York: Neale Publishing, 1911), 83–84.

2. Harry S. Arnold, "'Easy' Memories: The Way It Was," unpublished manuscript, 99th Infantry Division Collection, U.S. Army Military History Institute, Carlisle, Pennsylvania, 42 (hereafter cited as MHI).

3. Alvin Boeger Questionnaire, 99th Division Collection, MHI.

4. Martin van Creveld, *Fighting Power: German and U.S. Army Performance, 1939–1945* (Westport, CT: Greenwood Press, 1982), 170.

5. Anthony Kellett, *Combat Motivation: The Behavior of Soldiers in Battle* (Boston: Kluwer-Nijhoff, 1982), 334.

6. Paul Fussell, *Wartime: Understanding and Behavior in the Second World War* (New York: Oxford University Press, 1989), 282.

7. Jeffrey Carzales in Evan Wright, *Generation Kill: Devil Dogs, Captain America, and the New Face of American War* (New York: G. P. Putnam's Sons, 2004), 119.

8. Anthony Swofford, *Jarhead: A Marine's Chronicle of the Gulf War and Other Battles* (New York: Scribner, 2003), 195–196.

9. Brad Colbert in Wright, *Generation Kill*, 143.

10. Toby Winn in Trish Wood, *What Was Asked of Us: An Oral History of the Iraq War by the Soldiers Who Fought It* (New York: Little, Brown, 2006), 143.

11. Rupert Smith, *The Utility of Force: The Art of War in the Modern World* (New York: Alfred Knopf, 2007).

12. Garett Reppenhagen in Wood, *What Was Asked*, 187–188.

13. Wright, *Generation Kill*, 205.

14. Leonard Wong et al., *Why They Fight: Combat Motivation in the Iraq War* (Carlisle Barracks, PA: Strategic Studies Institute, U.S. Army War College, 2003), 23–25.

15. Ibid., 10–12.

16. Dominick King and Toby Winn in Wood, *What Was Asked of Us*, 161, 141.

17. Donovan Campbell, *Joker One: A Marine Platoon's Story of Courage, Leadership, and Brotherhood* (New York: Random House, 2009), 190.

18. Swofford, *Jarhead*, 123–124.

19. Milo S. Afong, *HOGs in the Shadows: Combat Stories from Marine Snipers in Iraq* (New York: Berkley Publishing Group, 2007), 2.

20. Antonio Espera in Wright, *Generation Kill*, 320.

21. Campbell, *Joker One*, 191.

22. Afong, *HOGs in the Shadows*, 4.

23. Christopher Lowman in William Finnegan, "The Last Tour," *New Yorker*, September 29, 2008, 68.

24. Campbell, *Joker One*, 190–191.

25. C. J. Chivers, "Foot on Bomb, A Marine Defies a Taliban Trap," *New York Times*, January 24, 2010, 1, 4.

26. Swofford, *Jarhead*, 159.

27. Samuel Stouffer et al., *The American Soldier* (Princeton, NJ: Princeton University Press, 1949), 2:90.

28. Richard A. Gabriel, *No More Heroes: Madness and Psychiatry in War* (New York: Hill and Wang, 1987), 130–151.

29. Eric T. Dean, *Shook over Hell: Post-Traumatic Stress, Vietnam, and the Civil War* (Cambridge, MA: Harvard University Press, 1997).

30. Fussell, *Wartime*, 282.

31. Arnold, "'Easy' Memories," 24.

32. Peter Richmond, *My Father's War: A Son's Journey* (New York: Simon & Schuster, 1996), 186.

Bibliography

Abbot, Henry Livermore. *Fallen Leaves: The Civil War Letters of Major Henry Livermore Abbot*. Ed. Robert Garth Shaw. Kent, OH: Kent State University Press, 1991.

Adkins, A. Z., Jr., and Andrew Z. Adkins, III. *You Can't Get Much Closer Than This: Combat with Company H, 317th Infantry Regiment, 80th Division*. Havertown, PA: Casemate Publishing, 2005.

Afong, Milo S. *HOGs in the Shadows: Combat Stories from Marine Snipers in Iraq*. New York: Berkley, 2007.

Allaire, Anthony. *Diary of Lieutenant Anthony Allaire*. New York: New York Times, 1968.

Ambrose, Stephen E. *Citizen Soldiers: The U.S. Army from the Normandy Beaches to the Bulge to the Surrender of Germany, June 7, 1944–May 7, 1945*. New York: Simon & Schuster, 1997.

Anderson, Fred. *A People's Army: Massachusetts Soldiers and Society in the Seven Years' War*. Chapel Hill: University of North Carolina Press, 1984.

Appy, Christian. *Working-Class War: American Combat Soldiers and Vietnam*. Chapel Hill: University of North Carolina Press, 1993.

Ardant du Picq, Charles. *Battle Studies: Ancient and Modern War*. Harrisburg, PA: Military Service Publishing, 1947.

Arn, Edward C. *Arn's War: Memoirs of a World War II Infantryman, 1940–1946*. Ed. Jerome Mushkat. Akron: University of Akron Press, 2006.

Astor, Gerald. *A Blood-Dimmed Tide: The Battle of the Bulge by the Men Who Fought It*. New York: Donald I. Fine, 1992.

———. *The Bloody Forest: The Battle for the Hürtgen, September 1944–January 1945*. Novato, CA: Presidio Press, 2000.

Atkins, Josiah. *The Diary of Josiah Atkins*. Ed. Steven E. Kagle. New York: New York Times, 1975.

Atkinson, Rick. *An Army at Dawn: The War in North Africa, 1942–1943*. New York: Henry Holt, 2002.

———. *Day of Battle: The War in Sicily and Italy, 1943–1944*. New York: Henry Holt, 2007.

———. *In the Company of Soldiers: A Chronicle of Combat*. New York: Henry Holt, 2005.

Atwell, Lester. *Private*. New York: Simon & Schuster, 1958.

Babcock, John B. *Taught to Kill: An American Boy's War from the Ardennes to Berlin*. Washington, DC: Potomac Books, 2005.

Babits, Lawrence E. *A Devil of a Whipping: The Battle of Cowpens*. Chapel Hill: University of North Carolina Press, 1998.

Bangs, Isaac. *Journal of Lieutenant Isaac Bangs*. Ed. by Edward Bangs. New York: New York Times, 1968.

Bardeen, Charles W. *A Little Fifer's War Diary*. Syracuse, NY: Printed by the author, 1910.

Bartov, Omer. "The Conduct of War: Soldiers and the Barbarization of Warfare." *Journal of Military History* 64 (December 1992): S32–S45.

———. *Hitler's Army: Soldiers, Nazis, and War in the Third Reich*. New York: Oxford University Press, 1991.

Baynes, John. *Morale: A Study of Men and Courage*. New York: Frederick A. Praeger, 1967.

Becker, Carl M., and Robert G. Theoben. *Common Warfare: Parallel Memoirs by Two World War II GIs in the Pacific*. Jefferson, NC: McFarland, 1992.

Bee, Robert L., ed. *The Boys from Rockville: Civil War Narratives of Sgt. Benjamin Hirst, Company D, 14th Connecticut Volunteers*. Knoxville: University of Tennessee Press, 1998.

Belenky, Gregory, ed. *Contemporary Studies in Combat Psychiatry*. New York: Greenwood Press, 1987.

Bernstein, Walter. *Keep Your Head Down*. New York: Viking, 1945.

Biddle, Stephen. *Military Power: Explaining Victory and Defeat in Modern Battle*. Princeton, NJ: Princeton University Press, 2004.

Bloomfield, Joseph. *Citizen Soldier: The Revolutionary War Journal of Joseph Bloomfield*. Ed. Mark E. Lender and James Kirby Martin. Newark: New Jersey Historical Society, 1982.

Blunt, Roscoe C., Jr. *Foot Soldier: A Combat Infantryman's War in Europe*. Cambridge, MA: Da Capo Press, 2001.

Bonner, Robert E., ed. *The Soldier's Pen: Firsthand Impressions of the Civil War*. New York: Hill and Wang, 2006.

Bourke, Joanna. *An Intimate History of Killing: Face-to-Face Killing in Twentieth-Century Warfare*. London: Granta Books, 1999.

Brewster, Charles Harvey. *When This Cruel War Is Over: The Civil War Letters of Charles Harvey Brewster*. Ed. David W. Blight. Amherst: University of Massachusetts Press, 1992.

Brown, David Tucker, Jr. *Letters of a Combat Marine*. Chapel Hill: University of North Carolina Press, 1947.

Bull, Rice C. *Soldiering: The Civil War Diary of Rice C. Bull, 123rd New York Volunteer Infantry*. Ed. Jack K. Bauer. San Rafael, CA: Presidio Press, 1977.

Burgett, Donald. *Currahee! A Screaming Eagle at Normandy*. Novato, CA: Presidio Press, 1999.

———. *Seven Roads to Hell: A Screaming Eagle at Bastogne*. Novato, CA: Presidio Press, 1999.

Byers, S. H. M. "How Men Feel in Battle: Recollections of a Private at Champion Hills." *Annals of Iowa* 2 (July 1896): 438–449.
———. *With Fire and Sword*. New York: Neale Publishing, 1911.
Callaway, Joshua K. *The Civil War Letters of Joshua K. Callaway*. Ed. Judith Lee Hallock. Athens: University of Georgia Press, 1997.
Cameron, Craig M. *American Samurai: Myth, Imagination, and the Conduct of Battle in the First Marine Division, 1941–1945*. Cambridge, UK: Cambridge University Press, 1994.
Campbell, Donovan. *Joker One: A Marine Platoon's Story of Courage, Leadership, and Brotherhood*. New York: Random House, 2009.
Carter, Robert Goldthwaite. *Four Brothers in Blue*. Austin: University of Texas Press, 1978.
Carter, Ross S. *Those Devils in Baggy Pants*. New York: Appleton-Century-Crofts, 1951.
Chamberlain, Joshua Lawrence. *Passing of the Armies: An Account of the Final Campaign of the Army of the Potomac Based Upon Personal Reminiscences of the Fifth Army Corps*. New York: G. P. Putnam's Sons, 1915.
Chisolm, Daniel. *The Civil War Notebook of Daniel Chisolm: A Chronicle of Daily Life in the Union Army, 1864–1865*. Ed. W. Springer Menge and J. August Shimrak. New York: Orion Books, 1989.
Colby, Elbridge. *Masters of Mobile Warfare*. Princeton, NJ: Princeton University Press, 1943.
Cole, Philip M. *Civil War Artillery at Gettysburg: Organization, Equipment, Ammunition and Tactics*. Cambridge, MA: Da Capo Press, 2002.
Connolly, James A. *Three Years in the Army of the Cumberland: The Letters and Diary of Major James A. Connolly*. Ed. Paul M. Angle. Bloomington: Indiana University Press, 1987.
Crane, Stephen. *The Red Badge of Courage*. London: Folio Society, 1951.
Crisp, Robert. *The Gods Were Neutral*. London: Frederick Muller, 1960.
Crow, Jeffrey, and Larry Tise, eds. *The Southern Experience in the American Revolution*. Chapel Hill: University of North Carolina Press, 1978.
Culver, J. F. *Your Affectionate Husband, J. F. Culver: Letters Written during the Civil War*. Ed. Leslie W. Dunlap. Iowa City: Friends of the University of Iowa Libraries, 1978.
Cunningham, Chet. *Hell Wouldn't Stop: An Oral History of the Battle of Wake Island*. New York: Carroll & Graf Publishers, 2002.
Dann, John C., ed. *The Revolution Remembered: Eyewitness Accounts of the War for Independence*. Chicago: University of Chicago Press, 1980.
Davidson, Bill. *Cut Off: Behind Enemy Lines in the Battle of the Bulge with Two Small Children, Ernest Hemingway, and Other Assorted Misanthropes*. New York: Stein and Day, 1972.
Dawes, Rufus R. *Service with the Sixth Wisconsin Volunteers*. Ed. Alan T. Nolan. Madison: State Historical Society of Wisconsin, 1962.
Dean, Eric T. *Shook over Hell: Post-Traumatic Stress, Vietnam, and the Civil War*. Cambridge: Harvard University Press, 1997.

DeForest, John William. *A Volunteer's Adventures: A Union Captain's Record of the Civil War.* Ed. James Croushore. New Haven, CT: Yale University Press, 1946.

Devitt, William. *Shavetail: The Odyssey of an Infantry Lieutenant in World War II.* St. Cloud, MN: North Star Press, 2001.

Dinter, Elmer. *Hero or Coward: Pressures Facing the Soldier in Battle.* Totowa, NJ: Frank Cass, 1985.

Dollard, John. *Fear in Battle.* New Haven, CT: Institute of Human Relations, 1943.

Dooley, John. *John Dooley, Confederate Soldier: His War Journal.* Ed. Joseph T. Durkin. Washington, DC: Georgetown University Press, 1945.

Doubler, Michael D. *Closing with the Enemy: How GIs Fought the War in Europe, 1944–1945.* Lawrence: University Press of Kansas, 1994.

Drez, Ronald J., ed. *Voices of D-Day: The Story of the Allied Invasion Told by Those Who Were There.* Baton Rouge: Louisiana State University Press, 1994.

DuPicq, Ardant. *Fighting Spirit.* Harrisburg, PA: Stackpole, 1958.

Dwight, Henry Otis. "How We Fight at Atlanta." *Harper's New Monthly Magazine* 29 (October 1864): 663–666.

Eaton, Joseph W. "Experiments in Testing for Leadership." *American Journal of Sociology* 52, no. 6 (May 1947): 523–535.

Edwards, Abial H. *Dear Friend Anna: The Civil War Letters of a Common Soldier from Maine.* Ed. Beverly Hayes Kallgren and James L. Crouthamel. Orono: University of Maine Press, 1992.

Evans, Thomas. "There Is No Use Trying to Dodge Shot." *Civil War Times Illustrated* 6 (January 1968): 45.

Fick, Nathaniel. *One Bullet Away: The Making of a Marine Officer.* Boston: Houghton Mifflin, 2005.

Field Manual 100-5. *Field Service Regulations.* Washington, DC: U.S. Government Printing Office, 1941.

Fisk, Wilbur. *Hard Marching Every Day: The Civil War Letters of Private Wilbur Fisk, 1861–1865.* Lawrence: University Press of Kansas, 1992.

Frank, Joseph Allan, and George Reaves. *"Seeing the Elephant": Raw Recruits at the Battle of Shiloh.* New York: Greenwood Press, 1989.

Frey, Sylvia R. *The British Soldier in North America: A Social History of Military Life in the Revolutionary Period.* Austin: University of Texas Press, 1981.

Fuller, J. F. C. *A Military History of the Western World.* Vol. 2. New York: Funk & Wagnalls, 1955.

Fussell, Paul. *Doing Battle: The Making of a Skeptic.* New York: Little, Brown, 1996.

———. *The Great War and Modern Memory.* New York: Oxford University Press, 1975.

———. *Wartime: Understanding and Behavior in the Second World War.* New York: Oxford University Press, 1989.

Gabel, Kurt. *The Making of a Paratrooper: Airborne Training and Combat in World War II.* Ed. William C. Mitchell. Lawrence: University Press of Kansas, 1990.

Gaff, Alan D. *On Many a Bloody Field: Four Years in the Iron Brigade.* Bloomington: Indiana University Press, 1996.

Gantz, Jacob. *Such Are the Trials: The Civil War Diaries of Jacob Gantz.* Ed. Kathleen Davis. Ames: Iowa State University Press, 1991.

Geer, Allen Morgan. *The Civil War Diary of Allen Morgan Geer.* Ed. Mary Ann Andersen. New York: Cosmos Press, 1977.

Gilbert, Benjamin. *Winding Down: The Revolutionary War Letters of Lieutenant Benjamin Gilbert of Massachusetts, 1780–1783.* Ed. John Shy. Ann Arbor: William Clements Library, 1989.

Glatthaar, Joseph T. *General Lee's Army: From Victory to Collapse.* New York: Free Press, 2008.

Gordon, Harold J., Jr. *One Man's War: A Memoir of World War II.* Ed. Nancy M. Gordon. New York: Apex Press, 1999.

Grant, Ulysses S. *Personal Memoirs of Ulysses S. Grant.* Cambridge, MA: Da Capo Press, 1982.

Gray, Jeffrey Alan. *The Psychology of Fear and Stress.* 2nd ed. Cambridge, UK: Cambridge University Press, 1987.

Gray, J. Glenn. *The Warriors: Reflections on Men in Battle.* New York: Harper & Row, 1970.

Greenman, Jeremiah. *Diary of a Common Soldier in the American Revolution, 1775–1783: An Annotated Edition of the Military Journal of Jeremiah Greenman.* Ed. Robert C. Bray and Paul E. Bushnell. DeKalb: Northern Illinois University Press, 1977.

Griffin, Richard N., ed. *Three Years a Soldier: The Diary and Newspaper Correspondence of Private George Perkins, Sixth New York Independent Battery, 1861–1864.* Knoxville: University of Tennessee Press, 2006.

Griffith, Paddy. *Rally Once Again: Battle Tactics of the American Civil War.* London: Crowood Press, 1987.

Grimsley, Mark. "In Not So Dubious Battle: The Motivations of American Civil War Soldiers." *Journal of Military History* 62, no. 1 (January 1998): 175–188.

Grinker, Roy R., and John P. Spiegel. *Men under Stress.* Philadelphia: Blakiston, 1945.

Grossman, David. *On Killing: The Psychological Cost of Learning to Kill in War and Society.* New York: Little, Brown, 1995.

Hannaford, Ebeneezer. "In the Ranks at Stones' River." In *Battles and Leaders of the Civil War,* ed. Peter Cozzens. Vol. 6. Champaign: University of Illinois Press, 2006.

Hardee, William J. *Rifle and Light Infantry Tactics.* Philadelphia: Lippincott, Grambo, 1855.

Hargrove, Marion. *See Here, Private Hargrove.* New York: Henry Holt, 1942.

Hastings, Max. *Armageddon: The Battle for Germany, 1944–1945.* New York: Knopf, 2004.

Haydon, Charles B. *For Country, Cause and Leader: The Civil War Journal of Charles B. Haydon.* Ed. Stephen W. Sears. Boston: Ticknor and Fields, 1993.

Henderson, William Darryl. *Cohesion: The Human Element in Combat: Leadership and Societal Influence in the Armies of the Soviet Union, the United States, North Vietnam, and Israel.* Washington, DC: National Defense University Press, 1988.

———. *Why the Vietcong Fought: A Study of Motivation and Control in a Modern Army in Combat.* Westport, CT: Greenwood Press, 1979.

Hess, Earl J. *The Union Soldier in Battle: Enduring the Ordeal of Combat.* Lawrence: University Press of Kansas, 1997.

Higginbotham, Don. *Daniel Morgan, Revolutionary Rifleman.* Chapel Hill: University of North Carolina Press, 1961.

———. *The War of American Independence: Military Attitudes, Policies, and Practice, 1763–1789.* New York: Macmillan, 1971.

Hocking, William Ernest. "The Nature of Morale." *American Journal of Sociology* 47, no. 3 (November 1941): 302–320.

Holmes, Oliver Wendell, Jr. *Touched with Fire: Civil War Letters and Diary of Oliver Wendell Holmes, Jr., 1861–1864.* Cambridge, MA: Harvard University Press, 1946.

Holmes, Richard. *Acts of War: The Behavior of Men in Battle.* New York: Free Press, 1985.

Hosea, Lewis M. "The Second Day at Shiloh." In *Sketches of War History, 1861–1865.* Ed. Theodore F. Allen, Edward S. McKee, and J. Gordon Taylor. Vol. 6. Cincinnati: R. Clark, 1908.

How to Get Along in the Army, by "Old Sarge." New York: D. Appleton-Century, 1942.

Hynes, Samuel. *The Growing Seasons: An American Boyhood before the War.* New York: Viking, 2003.

———. *The Soldiers' Tale: Bearing Witness to Modern War.* New York: Penguin Press, 1997.

Isham, A. B. "The Story of a Gunshot Wound." In *Sketches of War History, 1861–1865.* Ed. W. H. Chamberlin. Vol. 4. Cincinnati: R. Clark, 1896.

Johnson, Robert U., and C. C. Buel, eds. *Battles and Leaders of the Civil War.* New York: Century, 1884.

Jünger, Ernst. *Storm of Steel.* London: Chatto and Windus, 1929.

Keegan, John. *The Face of Battle.* New York: Viking, 1976.

Keiffer, Jesse. *The Civil War Letters and Diary of Jesse Kiefer of Lockport, New York.* Comp. Jeanne Kieffer and Craig Kieffer. Baltimore: Gateway Press, 2000.

Kellett, Anthony. *Combat Motivation: The Behavior of Soldiers in Battle.* Boston: Kluwer-Nijhoff, 1982.

Kennett, Lee. *G.I.: The American Soldier in World War II.* Norman: University of Oklahoma Press, 1987.

Ketchum, Richard L. *Saratoga: Turning Point of America's Revolutionary War.* New York: Henry Holt, 1997.

Kindsvatter, Peter S. *American Soldiers: Ground Combat in the World Wars, Korea, and Vietnam.* Lawrence: University Press of Kansas, 2003.

Kohn, Richard H. "The Social History of the American Soldier: A Review and

Prospectus for Research." *American Historical Review* 86, no. 3 (June 1981): 553–567.

Kviv, Frederick J. "Survival in Combat as a Collective Exchange Process." *Journal of Political and Military Sociology* 6 (1978): 219–232.

Lee, Wayne E. *Crowds and Soldiers in Revolutionary North Carolina: The Culture of Violence in Riot and War.* Gainesville: University of Florida Press, 2001.

Leed, Eric J. *No Man's Land: Combat and Identity in World War I.* New York: Cambridge University Press, 1979.

Lewis, Ralph. "Officer–Enlisted Men's Relationships." *American Journal of Sociology* 52, no. 5 (March 1947): 410–419.

Lind, Henry C., ed. *The Long Road Home: The Civil War Experiences of the Twenty-Seventh Massachusetts Regiment of Volunteer Infantry as Told by Their Personal Correspondence, 1861–1864.* Madison, NJ: Fairleigh Dickinson University Press, 1992.

Linderman, Gerald. *Embattled Courage: The Experience of Combat in the Civil War.* New York: Free Press, 1987.

———. *The World within War: America's Combat Experience in World War II.* New York: Free Press, 1997.

Little, Roger. "Buddy Relations and Combat Performance." In *The New Military: Changing Patterns of Organization,* ed. Morris Janowitz. New York: Russell Sage Foundation, 1964.

Lusk, William Thompson. *War Letters of William Thompson Lusk, Captain, Assistant Adjutant-General, United States Volunteers 1861–1863.* New York: Private printing, 1911.

Lynn, John A. *The Bayonets of the Republic: Motivation and Tactics in the Army of Revolutionary France, 1791–94.* Urbana: University of Illinois Press, 1984.

MacCoun, Robert. "What Is Known about Unit Cohesion and Military Performance." In *Sexual Orientation and U.S. Military Personnel Policy: Options and Assessment.* Santa Monica, CA: RAND, 1993.

MacCoun, Robert, Elizabeth Kier, and Aaron Belkin. "Does Social Cohesion Determine Motivation in Combat? An Old Question with an Old Answer." *Armed Forces and Society* 32 (2006): 646–654.

MacDonald, Charles B. *Company Commander.* Washington, DC: Infantry Journal Press, 1947.

Mackin, Elton E. *Suddenly We Didn't Want to Die: Memoirs of a World War I Marine.* Novato, CA: Presidio Press, 1993.

MacMillan, Harold. *Winds of Change, 1914–1939.* New York: Harper & Row, 1966.

Manchester, William. *Goodbye, Darkness: A Memoir of the Pacific War.* Boston: Little, Brown, 1979.

Manning, Chandra. *What This Cruel War Was Over: Soldiers, Slavery, and the Civil War.* New York: Alfred A. Knopf, 2007.

Mansoor, Peter R. *The GI Offensive in Europe: The Triumph of American Infantry Divisions, 1941–1945.* Lawrence: University Press of Kansas, 1999.

Marshall, S. L. A. *Battle at Best.* New York: William Morrow, 1963.

———. *Men against Fire: The Problem of Battle Command in Future War.* New York: William Morrow, 1947.

Maslowski, Peter. "A Study of Morale in Civil War Soldiers." *Military Affairs* 24 (1970): 122–126.

Matrau, Henry. *Letters Home: Henry Matrau of the Iron Brigade.* Ed. Marcia Reid-Green. Lincoln: University of Nebraska Press, 1993.

Mauldin, Bill. *Bill Mauldin's Army.* Novato, CA: Presidio Press, 1979.

McDonough, James Lee. *Shiloh: In Hell before Night.* Knoxville: University of Tennessee Press, 1977.

McMahon, John T. *John T. McMahon's Diary of the 136th New York, 1861–1864.* Ed. John Michael Priest. Shippensburg, PA: White Mane, 1993.

McManus, John C. *The Deadly Brotherhood: The American Combat Soldier in World War II.* Novato, CA: Presidio Press, 1998.

McNeill, William H. *Keeping Together in Time: Dance and Drill in Human History.* Cambridge, MA: Harvard University Press, 1995.

McPherson, James. *Battle Cry of Freedom: The Civil War Era.* New York: Oxford University Press, 1988.

———. *For Cause and Comrades: Why Men Fought in the Civil War.* New York: Oxford University Press, 1997.

———. *What They Fought For, 1861–1865.* Baton Rouge: Louisiana State University Press, 1994.

McWhiney, Grady, and Perry Jamieson. *Attack and Die: Civil War Military Tactics and the Southern Heritage.* Tuscaloosa: University of Alabama Press, 1982.

Middlekauff, Robert. "Why Men Fought in the American Revolution." *Huntington Library Quarterly* 43 (Spring 1980): 135–148.

Miller, Edward G. *A Dark and Bloody Ground: The Hürtgen Forest and the Roer River Dams, 1944–1945.* College Station: Texas A&M University Press, 1995.

Miller, William Ian. *The Mystery of Courage.* Cambridge, MA: Harvard University Press, 2000.

Mitchell, Reid. *Civil War Soldiers: Their Expectations and Their Experiences.* New York: Viking Press, 1980.

———. *The Vacant Chair: The Northern Soldier Leaves Home.* Oxford: Oxford University Press, 1993.

Moran, Charles. *Anatomy of Courage.* Garden City Park, NY: Avery, 1987.

Moser, Richard R. *The New Winter Soldiers: GI and Veteran Dissent during the Vietnam Era.* New Brunswick, NJ: Rutgers University Press, 1996.

Moskos, Charles C., Jr. "The American Combat Soldier in Vietnam." *Journal of Social Issues* 31 (1975): 25–37.

———. *The American Enlisted Man: The Rank and File in Today's Military.* New York: Russell Sage Foundation, 1970.

Mosman, Chesley A. *The Rough Side of War: The Civil War Journal of Chesley A. Mosman.* Ed. Arnold Gates. Garden City, NY: Basin, 1987.

Moss, James A. *Officers' Manual.* Menasha, WI: George Banta, 1917.

Mumford, Lewis. *Green Memories: The Story of Geddes Mumford.* New York: Harcourt, Brace, 1947.

Munson, Edward Lyman, Jr. *Leadership for American Army Leaders.* Washington, DC: Infantry Journal, 1942.

Musser, Charles O. *Soldier Boy: The Civil War Letters of Charles O. Musser, 29th Iowa.* Ed. Barry Popchock. Iowa City: University of Iowa Press, 1995.

Neill, George. *Infantry Soldier: Holding the Line at the Battle of the Bulge.* Norman: University of Oklahoma Press, 2000.

Newton, James K. *A Wisconsin Boy in Dixie.* Madison: University of Wisconsin Press, 1961.

Perry, David. "Recollections of an Old Soldier." *Magazine of History* 137 (1928).

Powell, William S. "A Connecticut Soldier under Washington: Elisha Bostwick's Memoirs of the First Years of the Rebellion." *William and Mary Quarterly,* 3rd ser., 6 (1949): 94–107.

Power, J. Tracy. "'The Virtue of Humanity Was Totally Forgot': Buford's Massacre, May 29, 1780." *South Carolina Historical Magazine* 93, no. 1 (January 1992): 5–14.

Psychology for the Fighting Man, Prepared for the Fighting Man Himself by a Committee of the National Research Council with the Collaboration of Science Service as a Contribution to the War Effort. Washington, DC: Infantry Journal, 1943.

Pugh, Robert C. "The Revolutionary Militia in the Southern Campaign, 1780–1781." *William and Mary Quarterly* 14, no. 2 (April 1957): 154–175.

Pyle, Ernie. *Here Is Your War.* New York: Henry Holt, 1943.

———. *Last Chapter.* New York: Henry Holt, 1946.

Rachman, S. J. *Fear and Courage.* San Francisco: W. H. Freeman, 1978.

Reed, Charles Wellington. *"A Grand Terrible Dramma" from Gettysburg to Petersburg: The Civil War Letters of Charles Wellingon Reed.* Ed. Eric A. Campbell. New York: Fordham University Press, 2000.

Reyburn, Phillip J., and Terry L. Wilson, eds. *"Jottings from Dixie": The Civil War Dispatches of Sergeant Major Stephen F. Flaherty, U.S.A.* Baton Rouge: Louisiana State University Press, 1999.

Rhea, Gordon C. *The Battle of the Wilderness, May 5–6, 1864.* Baton Rouge: Louisiana State University Press, 1994.

Richardson, F. M. *Fighting Spirit: A Study of Psychological Factors in War.* London: Leo Cooper, 1978.

Richmond, Peter. *My Father's War: A Son's Journey.* New York: Simon & Schuster, 1996.

Robertson, James I., Jr. *Soldiers Blue and Gray.* Columbia: University of South Carolina Press, 1988.

Rodgers, Thomas E. "Billy Yank and G.I. Joe: An Exploratory Essay on the Sociopolitical Dimensions of Soldier Motivation." *Journal of Military History* 69 (January 2005): 93–121.

Ross, Steven T. *From Flintlock to Rifle: Infantry Tactics, 1740–1866.* London: Frank Cass, 1979.

Royster, Charles. *A Revolutionary People at War: The Continental Army and American Character, 1775–1783.* Chapel Hill: University of North Carolina Press, 1979.

Rush, Robert S. *Hell in Hürtgen Forest: The Ordeal and Triumph of an American Infantry Regiment.* Lawrence: University Press of Kansas, 2001.

Russ, Martin. *The Last Parallel: A Marine's War Journal*. New York: Rinehart, 1957.

Ryan, Dennis P., ed. *A Salute to Courage: The American Revolution as Seen through the Wartime Writings of Officers of the Continental Army and Navy*. New York: Columbia University Press, 1979.

Sarkesian, Sam, ed. *Combat Effectiveness: Cohesion, Stress, and the Volunteer Military*. Beverly Hills, CA: Sage, 1980.

Seligman, Martin. *Learned Helplessness: A Theory for the Age of Personal Control*. New York: Oxford University Press, 1993.

Shanklin, James Maynard. *Dearest Lizzie: The Civil War Letters of Lt. Col. James Maynard Shanklin*. Ed. Kenneth P. McCutchan. Evansville, IN: Friends of Willard Library Press, 1988.

Shay, Jonathan. *Achilles in Vietnam: Combat Trauma and the Undoing of Character*. New York: Simon & Schuster, 1994.

Shephard, Ben. *A War of Nerves: Soldiers and Psychiatrists in the Twentieth Century*. Cambridge, MA: Harvard University Press, 2001.

Shils, Edmund, and Morris Janowitz. "Cohesion and Disintegration in the Wehrmacht in World War II." *Public Opinion Quarterly* 12, no. 2 (Summer 1948): 280–315.

Skelton, William B. "High Army Leadership in the Era of the War of 1812: The Making and Remaking of the Officer Corps." *William and Mary Quarterly* 51, no. 2 (April 1994): 253–274.

Sledge, Eugene B. *With the Old Breed at Peleliu and Okinawa*. Novato, CA: Presidio Press, 1981.

Slim, William. *Defeat into Victory*. New York: David McKay, 1961.

Small, Abner. *The Road to Richmond: The Civil War Memoirs of Major Abner R. Small*. Ed. Harold A. Small. Berkeley: University of California Press, 1959.

Smith, Rupert. *The Utility of Force: The Art of War in the Modern World*. New York: Alfred Knopf, 2007.

Solomon, Zahava. *Combat Stress Reaction: The Enduring Toll of War*. New York: Plenum Press, 1993.

Spiller, Roger J. "S. L. A. Marshall and the Ratio of Fire." *RUSI Journal* 133, no. 4 (Winter 1988): 63–71.

Standifer, Leon C. *Not in Vain: A Rifleman Remembers World War II*. Baton Rouge: Louisiana State University Press, 1992.

Stannard, Richard M. *Infantry: An Oral History of a World War II American Infantry Battalion*. New York: Twayne, 1993.

Steuben, Frederick William. *Regulations for the Order and Discipline of the Troops of the United States*. Philadelphia: Styner & Cist, 1779.

Stewart, Nora Kinzer. *Mates and Muchachos: Unit Cohesion in the Falklands/Malvinas War*. New York: Brassey's (U.S.), 1991.

Stillwell, Leander. *The Story of a Common Soldier of Army Life in the Civil War, 1861–1865*. Kansas City, MO: Franklin Hudson, 1920.

Stouffer, Samuel, et al. *The American Soldier*. Vol. 2, *Combat and Its Aftermath*. Princeton, NJ: Princeton University Press, 1949.

Strachan, Hew. "Training, Morale and Modern War." *Journal of Contemporary History* 42, no. 2 (2006): 211–227.

Strother, David Hunter. *A Virginia Yankee in the Civil War: The Diaries of David Hunter Strother.* Ed. Cecil D. Eby, Jr. Chapel Hill: University of North Carolina Press, 1961.

Swofford, Anthony. *Jarhead: A Marine's Chronicle of the Gulf War and Other Battles.* New York: Scribner, 2003.

Thomas, James B. *"I Never Again Want to See Such Sights": The Civil War Letters of First Lieutenant James B. Thomas.* Ed. Mary Warner Thomas and Richard A. Sauers. Baltimore: Butternut and Blue, 1995.

Travers, Tim. *The Killing Ground: The British Army, the Western Front, and the Emergence of Modern Warfare 1900–1918.* London: Allen & Unwin, 1987.

U.S. Army Surgeon General's Office. *The Medical and Surgical History of the War of the Rebellion, 1861–1865.* Washington, DC: U.S. Government Printing Office, 1870.

Van Creveld, Martin. *Fighting Power: German and U.S. Army Performance, 1939–1945.* Westport, CT: Greenwood Press, 1982.

Vanderhoef, Lorenzo. *"I Am Now a Soldier!": The Civil War Diaries of Lorenzo Vanderhoef.* Ed. Kenneth R. Martin and Ralph Linwood Snow. Bath, ME: Patten Free Library, 1990.

Waid, Seth, III. *The Civil War Diaries of Seth Waid III.* Ed. Robert Ilesevich and Jonathan Helmreich. Meadville, PA: Crawford County Historical Society, 1993.

Walker, Aldace Freeman. *Quite Ready to be Sent Somewhere: The Civil War Letters of Aldace Freeman Walker.* Ed. Tom Ledoux. Victoria, BC: Trafford, 2002.

Watkins, Sam R. *"Co. Aytch": A Side Show of the Big Show.* New York: Macmillan, 1962.

Watson, Peter. *War on the Mind: The Uses and Abuses of Psychology.* New York: Basic Books, 1978.

Weigley, Russell. *History of the United States Army.* New York: Macmillan, 1967.

———. *The Partisan War: The South Carolina Campaign of 1780–1782.* Columbia: University of South Carolina Press, 1970.

Weinberg, Gerhard. *A World at Arms: A Global History of World War II.* Cambridge, UK: Cambridge University Press, 1994.

Weitz, Mark A. "Drill, Training, and the Combat Performance of the Civil War Soldier: Dispelling the Myth of the Poor Soldier, Great Fighter." *Journal of Military History* 62, no. 2 (April 1998): 263–289.

Weller, Jac. "The Irregular War in the South." *Military Affairs* 24, no. 3 (autumn 1960): 124–136.

Welsh, Peter. *Irish Green and Union Blue: The Civil War Letters of Peter Welsh, Color Sergeant, Twenty-Eighth Regiment, Massachusetts Volunteers.* Ed. Lawrence Frederick Cole and Margaret Cosse Richard. New York: Fordham University Press, 1986.

Wheeler, Richard, ed. *Voices of the Civil War.* New York: Thomas Y. Crowell, 1976.

Whitson, Jennifer A., and Adam D. Galinsky. "Lacking Control Increases Illusory Pattern Perception." *Science* 322, no. 115 (October 2008): 115–117.

Wiley, Bell I. *The Life of Billy Yank, the Common Soldier of the Union.* New York: Bobbs Merrill, 1951.

———. *The Life of Johnny Reb, the Common Soldier of the Confederacy.* New York: Bobbs Merrill, 1943.

Williams, Alpheus S. *From the Cannon's Mouth.* Ed. Milo P. Quaife. Detroit: Wayne State University Press, 1959.

Williams, Robin M., Jr. "Field Observations and Surveys in Combat Zones." *Social Psychology Quarterly* 47, no. 2 (1987): 186–192.

Wilson, Charles. *The Anatomy of Courage: The Classic World War I Study of the Psychological Effects of War.* London: Constable, 1945.

Wong, Leonard, et al. *Why They Fight: Combat Motivation in the Iraq War.* Carlisle Barracks, PA: Strategic Studies Institute, U.S. Army War College, 2003.

Wood, Trish. *What Was Asked of Us: An Oral History of the Iraq War by the Soldiers Who Fought It.* New York: Little, Brown, 2006.

Wright, Catherine M., ed. *Lee's Last Casualty: The Life and Letters of Sgt. Robert W. Parker, Second Virginia Cavalry.* Knoxville: University of Tennessee Press, 2008.

Wright, Evan. *Generation Kill: Devil Dogs, Captain America, and the New Face of American War.* New York: G. P. Putnam's Sons, 2004.

Young, Thomas. "Memoir of Major Thomas Young." *Orion Magazine* 3 (October/November 1843).

Zimbardo, Philip. *The Lucifer Effect: Understanding How Good People Turn Evil.* New York: Random House, 2007.

Index

support troops, 197, 254n68
survival. *See* self-preservation
Swofford, Anthony, 21, 213–214, 217

tactics
  artillery, 50
  German, 44–45, 119
  move-and-fire, 45–46, 186
  See also dispersed battlefield; linear
    battlefield
tanks, 53–54, 155, 161, 162, 164
task cohesion, 177
technological change
  in ammunition, 152, 157
  effects on battlefield experiences, 11,
    55–57
  motivating factors and, 203–204
  tactical changes driven by, 7–9
  in training, 215
  in twenty-first century, 208
  in weaponry, 7–8, 152, 154–155, 160
thirst. *See* dehydration
Thompson, David, 1, 65
Tillotson, George, 74
tombstones, mock, 113, 216
training
  basic, 95–96, 103–104, 113, 114–115,
    119, 145
  battlefield simulations, 97, 103, 107–
    108, 110–111, 114–115, 119, 205, 215
  benefits, 106–107
  casualties in, 111, 115
  changes over time, 95–98, 103, 108,
    205, 239n6
  in Civil War, 95, 107, 109
  commonalities over time, 104–107
  in Continental Army, 95, 96, 97,
    221–222n6
  coordination of different units,
    116–117
  differences from combat, 118–120, 215
  for dispersed battlefield, 12, 103–120
  drills, 95, 97, 98, 99–101, 104–105,
    106–107, 208
  effectiveness, 101, 117–118
  engagement in, 113, 115
  goals, 96, 99–100, 103

identity shifts in, 99, 103–104
instructors, 109–110, 116, 198,
  241n36
lectures, 115, 141–142, 145
lessons of historic battles, 21
for linear battlefield, 95, 96–98,
  99–103
live ammunition in, 241n41
manuals of arms, 95, 104–106, 159
of officers, 40–41, 53, 110, 116, 123,
  133–134, 137, 141–142
overlearning tasks, 101–102, 105–106,
  110, 208, 240n16
preparation for killing, 225n20,
  238–239n5
repetition, 95, 105
rote memorization, 100, 101–102
shortcomings, 118–120
in skills to stay alive, 15–16, 93, 108,
  111, 113–115, 170, 214, 216
socialization, 96, 99
specialized, 104
stages, 103–104
target practice, 108–109, 159, 213–214
in twenty-first century, 208, 215, 216
weapons, 101–102, 104–106, 108–109,
  110, 155–156
in World War II, 95–98, 99, 103–120
training manuals, 95, 97, 98, 105,
  111–113
Transportation Corps, 197, 254n68
tree bursts, 50–51
trenches, 46, 51–52, 74
trophies, 220
truck drivers, 197, 254n68
trust
  on dispersed battlefield, 50, 180,
    191–192, 197–198
  on linear battlefield, 179–184
Twelve O'Clock High, 250–251n16

unconventional warfare, 209–211,
  215–216, 218
uniforms
  in Civil War, 40, 75
  of Continental Army, 75
  in World War II, 186